MS GULF COAST COMMUNITY COLLEGE
JEFFERSON DAVIS CAMPUS LIBRARY

Plankton
A Microscopic World

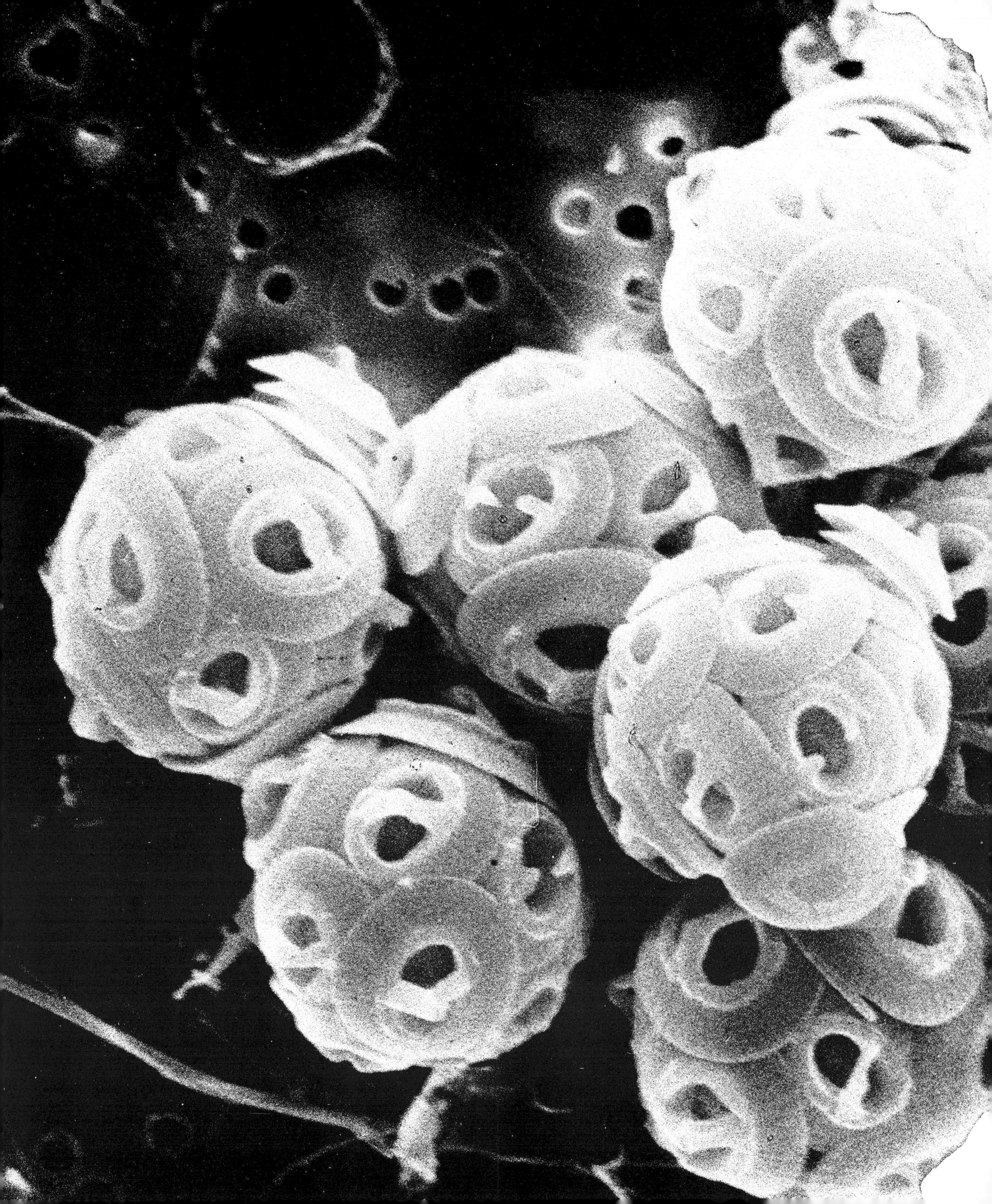

Plankton
A Microscopic World

Gustaaf M. Hallegraeff

E.J. BRILL
LEIDEN · NEW YORK · KØBENHAVN · KÖLN
1988

Designed and produced by
E.J. Brill/Robert Brown and Associates
This edition published
and distributed world wide by
E.J. Brill Publishing Company
P.O. Box 9000
NL-2300 PA Leiden
The Netherlands

USA and Canada
E.J. Brill (USA) Inc
1780 Broadway
Suite 1004
New York NY 10010

Cataloging-in-Publication entry

Hallegraeff, G.M. (Gustaaf M.).
Plankton: a miscroscopic world.
Bibliography.
Includes index.
ISBN 90 04 08932 2
1. Phytoplankton - Pictorial works.
2. Plankton - Pictorial works. I. Title.

589.4'022'2

Printed in Hong Kong

10 9 8 7 6 5 4 3 2 1

Contents

SECTION 1

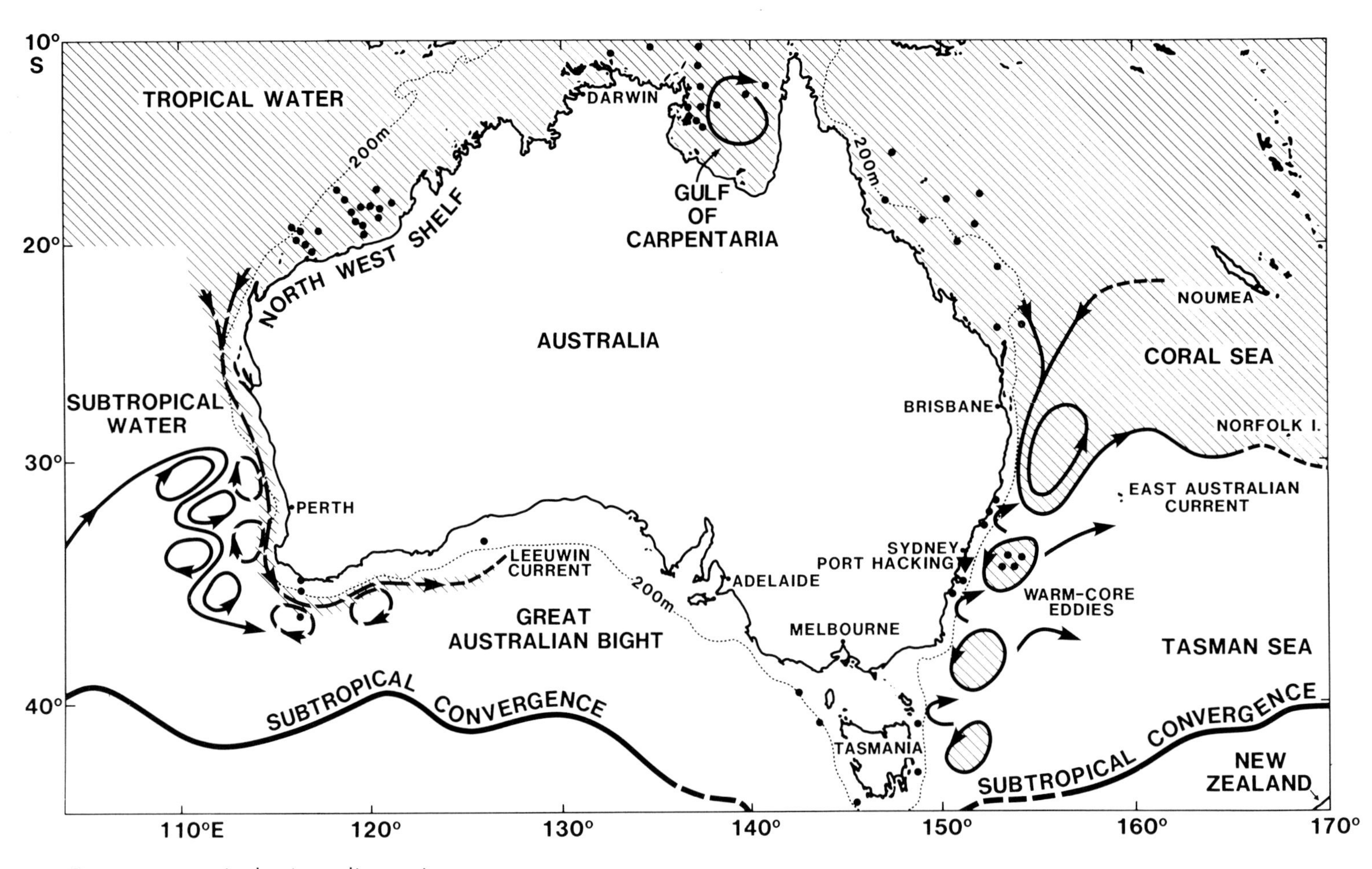

Ocean currents in the Australian region.

Introduction

Imagine being a space traveller who visits the planet Earth some 400 million years ago. Macroscopic plant life had not yet evolved, and to the unaided human eye this planet would have appeared barren and inhospitable. However, if our visitor were equipped with a microscope,he or she would have been bedazzled to discover the myriad of microscopic, one-celled life forms floating in the rich biochemical soup of this planet's bodies of water. Unperturbed by the fantastic permutations of multi-cellular life forms which have since developed through the millenia, descendants of these first cells still populate the present-day oceans.

WHAT IS PLANKTON?

On land, plants are conspicuous everywhere, whereas in the sea the only plants visible to the casual observer are the seaweeds along our rocky shores and the seagrasses of shallow estuaries. The enormous productivity of the oceans, which cover 70% of the Earth's surface (our planet should more appropriately have been called Water!), is based on untold millions of unicellular microscopic algae, collectively called *phytoplankton* ('phyto' = plant; 'planktos' = made to wander). These minute plants range in size from 1/1000 of a millimetre to 2 millimetres and live a floating existence in the upper 200 m of the ocean, where sunlight is available for photosynthesis.

All oxygen-breathing creatures, including humans, are indebted to the phytoplankton because through millenia of photosynthesis they have contributed significantly to the oxygen we breathe. Furthermore, much of the oil that we use today probably began millions of years ago when the sun shone on plankton drifting in prehistoric seas and produced, through photosynthesis, minute globules of oil within these cells.

Some microscopic fossils from sediments are remarkably similar to living microscopic plankton, indicating that some species have remained virtually unchanged for millions of years. Because of these creatures' small size, their unique shapes and structures have only become appreciated in the past two decades through the development of electron microscopy.

MICROSCOPES USING ELECTRONS INSTEAD OF LIGHT

In the 17th century, the Dutch pioneer microscopist Anton van Leeuwenhoek, using crude glass lenses, was probably the first human being to see minute creatures, which he called 'animalcules', in pond water. In the 1950s an instrument was developed, called the electron microscope, which allowed scientists their first close-up look of this architecturally interesting and aesthetically beautiful microscopic world.

Unlike an optical microscope, this $100,000 instrument uses no beams of light and no glass lenses. Instead, it bombards biological specimens with a thousand-volt beam of electrons, focussed and manipulated by magnetic lenses. The best optical microscopes, limited by the wavelengths of light , magnify no more than 1,000 times and cannot resolve, or discriminate, objects closer together than about 0.2 microns (1 micron = 1/1000 of a millimetre). Instead the electron microscope uses an electron beam with a wavelength about 10,000 times less than the average wavelength of light, and this offers an increase in resolving power to about 0.01 micron. In the scanning electron microscope (SEM) a beam of electrons scans the outer surface of objects previously coated with a thin layer of gold or platinum, whereas in the transmission electron microscope (TEM) electrons pass through thin biological specimens. Most illustrations in this book have been prepared with the scanning electron microscope, but a few cells have been photographed using both instruments to show further fine structure (compare pages 27 and 28, 33 and 34, 35 and 36).Useful magnifications range from 20 times to 30,000 times for scanning electron microscopy and up to 100,000 times for transmission electron microscopy. The scanning electron microscope images are of particular value to study plankton because of their great depth of focus, unachievable with conventional optical microscopes. Unlike light rays, electrons cannot convey colours, which is why all the illustrations in this book are in black and white. The scanning electron micrographs in this book were taken with a JEOL-35C or Philips-515 microscope using Ilford FP4 or Agfapan 25 film. The transmission electron micrographs were taken with a Philips-400 or Philips-410 instrument, using Kodak 4489 film. Prints and enlargements were made on Ilford Multigrade photographic paper.

This book presents but a small selection from a collection of over ten thousand electron micrographs that we have accumulated in the course of our research in Australian waters over the past ten years. I hope you will enjoy these micrographs as beautiful art forms, as a source of inspiration for marine biologists or perhaps architects, or simply to appreciate the diversity of life forms found on our unique planet.

DIVERSITY OF UNICELLULAR LIFE FORMS

A single litre of seawater may contain as many as 1 million microscopic plant cells from up to sixty different species. Unlike plant life on the land which is dominated by a single class of organism (the 'higher plants'), plant life in the oceans includes representatives of as many as thirteen algal classes. These organisms show an immense diversity of form, pigmentation and cellular structure, which are all adaptations to the oceanic environment. Phytoplankton species range from the primitive blue-green algae (sometimes called cyanobacteria) which were among the first living organisms on our planet, through the various golden-brown algal groups such as diatoms, dinoflagellates and coccolithophorids, to the green flagellates which are thought to have been the immediate precursors of the higher plants on the land.
This book illustrates almost exclusively species that have thick outer cell walls, such as the *coccolithophorids* (with calcareous scales), *diatoms* (with silica walls) and *dinoflagellates* (with cellulose armour). A few microscopic plankton animals (collectively called *zooplankton*) such as tintinnids, foraminiferans and radiolarians are also included (pages 100 to 105). It must be noted that there are also hundreds of so-called 'naked' plankton species, which are bounded only by a membranous covering (pages 50 to 51). These organisms are very easily damaged and are difficult to study by any form of microscopy. Scientific names for all the species photographed are provided as a reference for marine biologists, but also because the Latin etymology of these names often reflects aesthetic qualities of the organisms.

COCCOLITHOPHORIDS

The coccolithophorids ('coccus' = berry; 'lithos' = stone; 'phorid' = carrying) are a class of small golden-brown coloured flagellates, mostly 2 to 20 microns in size, which are covered by delicate calcareous scales known as coccoliths (pages 6 to 16). Composed of calcium carbonate crystals, these coccoliths form inside the cell, before being transported to the external surface of the cell where they are deposited. Nobody really knows what value these delicate body scales are to the algal cells. Candidate theories include a skeleton function, easily discarded ballast for helping the organism rise through the water, a light-scattering function to stimulate photosynthesis, or simply God's gift to taxonomists.
About 300 species of coccolithophorids are known in our present day oceans. These organisms are also known from sedimentary deposits dating back to Jurassic times (over 150 million years ago), but they were most abundant in the Cretaceous period, which is named after the extensive calcareous deposits of coccolithophorid origin in this era. The white cliffs of Dover, England, are an example of this kind of deposit. In recent years dense concentrations of coccolithophorids (up to 1 million cells per litre), in, for example Norwegian waters, have occasionally resulted in milky-white seas. Coccoliths were first discovered in marine sediments as long ago as 1836, but at that time they were thought to be crystals of inorganic origin. Only by the turn of the century were living coccolithophorids discovered swimming freely in the surface layers of the ocean, and was their biological origin recognized. Studies of the distribution of coccolithophorids in marine sediments are important in the field of oil exploration.

DIATOMS

The silica-walled diatoms ('diatoma' = cut in half) are perhaps the best known and most abundant of all the unicellular plankton algae. Their pill-box shaped silica shells, known as frustules, (pages 17 to 48) range in size from 2 to 2000 micron in diameter and are exquisitely ornamented with pores, ribs and spines radiating in spectacular geometry across the minute cell surfaces. These tiny plants can occur as single cells or in long chains with the cells linked by mucus threads or interlocking spines. Valve symmetry is used to distinguish centric diatoms with a radial symmetry and pennate diatoms with a symmetry around a line. Diatom frustules have fascinated microscopists ever since they were first discovered by Van Leeuwenhoek in 1702. In the past two decades, the application of electron microscopes has allowed scientists to identify many new features of diatom cell walls. The variety of the silica wall, upon which diatom taxonomy is based, has resulted in more than 5,000 species being described.
Diatoms are found floating in the plankton of the oceans, but also in the plankton of lakes and estuaries, especially when nutrients are brought to the surface by winds and currents. Plankton diatoms are the main food for small vegetarian animals, such as copepods and shrimplike krill, which in turn are eaten by small fish. Other diatoms live on the bottom of shallow estuaries, accounting for the brown film on tidal mudflats. Just after the tide goes out, these mudflat diatoms glide to the surface for a sunbath. They know precisely when to burrow back into the mud before the return of the tide. Diatoms also form a brown coating under polar ice floes or on whale skins. On land, they live under moist conditions in top soil, attached to moss, tree trunks or even brick walls.
The fossil record of centric diatoms begins in the Jurassic (150 million years ago) whereas pennate diatoms are known since the Tertiary (60 million years ago). Fossil diatom shells can carpet the seafloor with a layer of diatomaceous ooze as deep as 300 metres. This material has industrial applications in filters, as a fine abrasive, a filler for paint and is used in many other products.

DINOFLAGELLATES

Dinoflagellates ('dino' = whirling) have whip-like flagella that help them move up or down in the water column. They are a remarkably diverse group (pages 49 to 98), comprising over 2,000 living species that range in size from 2 to 2000 micron. About half of the dinoflagellate species behave as photosynthetic plants, while others behave more like animals, grazing on smaller cells or absorbing dissolved organic matter.

Dinoflagellate cells are often encased by an armour of cellulose plates that fit together like jigsaw pieces. The number, shape and arrangement of these plates are important for species identification. Scanning electron microscopy has revealed that ornamentation of the thecal plates and fine structure of the apical pore (page 84), present on top of many cells are also useful diagnostic characters. Dinoflagellates, like diatoms, can exhibit extensive morphological modifications of spines, horns and wing-like structures, especially in tropical species. These morphological extensions are thought to aid flotation, protect against grazing by small zooplankton, and enhance nutrient uptake through increased surface area and rotational movements.

Dinoflagellates are an important component of the plankton of marine and to lesser extent, freshwater environments, and they also play a key role as the symbiotic 'zooxanthellae' of coral reef animals. Some species have attracted considerable attention because of their bioluminescent properties, which can light up the sea at night. Others produce red tides. Red tides are blooms of dinoflagellates that grow so dense that they colour the sea red (which is how the Red Sea got its name). These plankton blooms can kill fish and other marine fauna, either through the generation of anoxic conditions in sheltered bays or more seriously by producing potent neurotoxins that can find their way through the food chain to humans. The red tide dinoflagellate *Pyrodinium bahamense* (pages 75 to 77), for example, has caused fatal cases of paralytic shellfish poisoning in Papua New Guinea after people ate contaminated shellfish. The bottom-dwelling dinoflagellate *Gambierdiscus toxicus* (pages 83 and 84) is responsible for ciguatera fish poisoning in tropical seas such as the Great Barrier Reef region.

Some present-day dinoflagellates form a resistant stage, or cyst, in their life cycle, which is generally produced at the onset of adverse conditions. Cysts settle to the bottom of the sea or lake and, when conditions are suitable, may germinate to the normal plankton stage. These cysts resemble fossilised 'dinocysts', which are abundant in sedimentary deposits at least as far back as Jurassic times.

PLANKTON FLORA OF AUSTRALIAN WATERS

In contrast to its unique terrestrial flora, because of the continuity of the oceans, the marine plankton flora of Australia was not isolated and shows broad similarities to that of other parts of the world. Warm- and cold-water plankton floras can be distinguished. These environments have different species of large diatoms and dinoflagellates whereas smaller species (such as coccolithophorids) are remarkably similar in all environments. Tropical diatoms and dinoflagellates of the North West Shelf of Australia, the Gulf of Carpentaria and Coral Sea have long hairs and wings that help them stay afloat in the less viscous warm tropical waters. The sporadic occurrence of tropical species in the cooler Tasman Sea and Great Australian Bight can be used as 'fingerprints' for the southwardly flowing East Australian Current and Leeuwin Current, respectively. Cold-water species from New South Wales and Tasmanian coastal waters, on the other hand, are similar to species from northern temperate seas.

Scientists study the factors that influence the growth, distribution and species succession of these minute plants and try to find out which species are critical food for the larvae of commercially important fish. Failure of a particular plankton species to occur in certain cell densities at the right place and the right time of the year can mean failure of an entire year class of a fishery!

SECTION 2

Coccolithophorid *Discosphaera tubifera* with trumpet-shaped appendices; Sydney coastal waters; diameter 15 μm.

Tropical coccolithophorid *Scyphosphaera apsteinii* carrying disc-shaped and vase-shaped coccoliths on one and the same cell. The sporadic occurrence of this warm-water species in the cooler Tasman Sea and Great Australian Bight reflects southward transport by the East Australian Current and Leeuwin Current, respectively; diameter 40 μm.

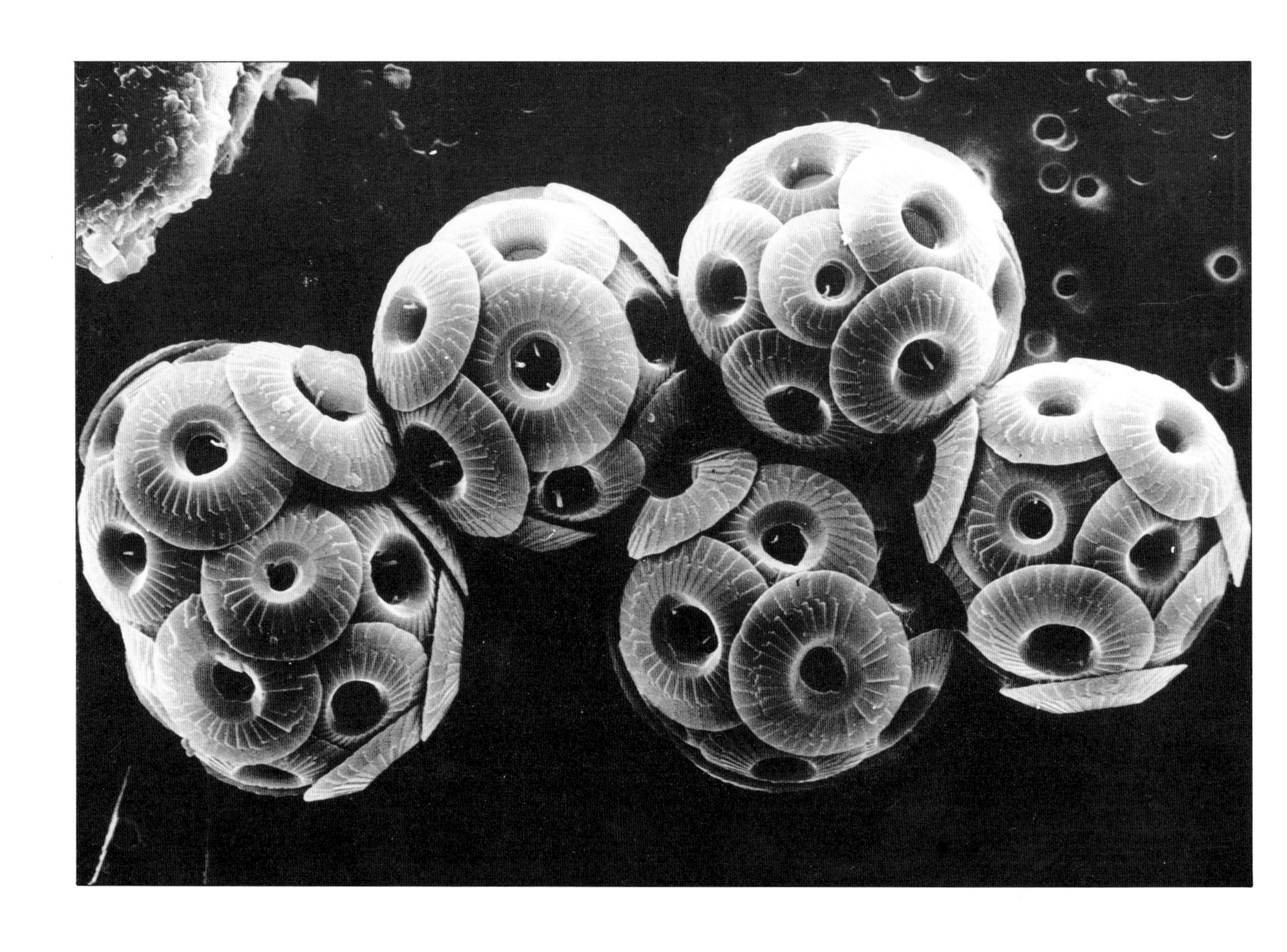

Cluster of the coccolithophorid *Umbilicosphaera sibogae*; North West Shelf of Australia; individual cells 20μm in diameter.

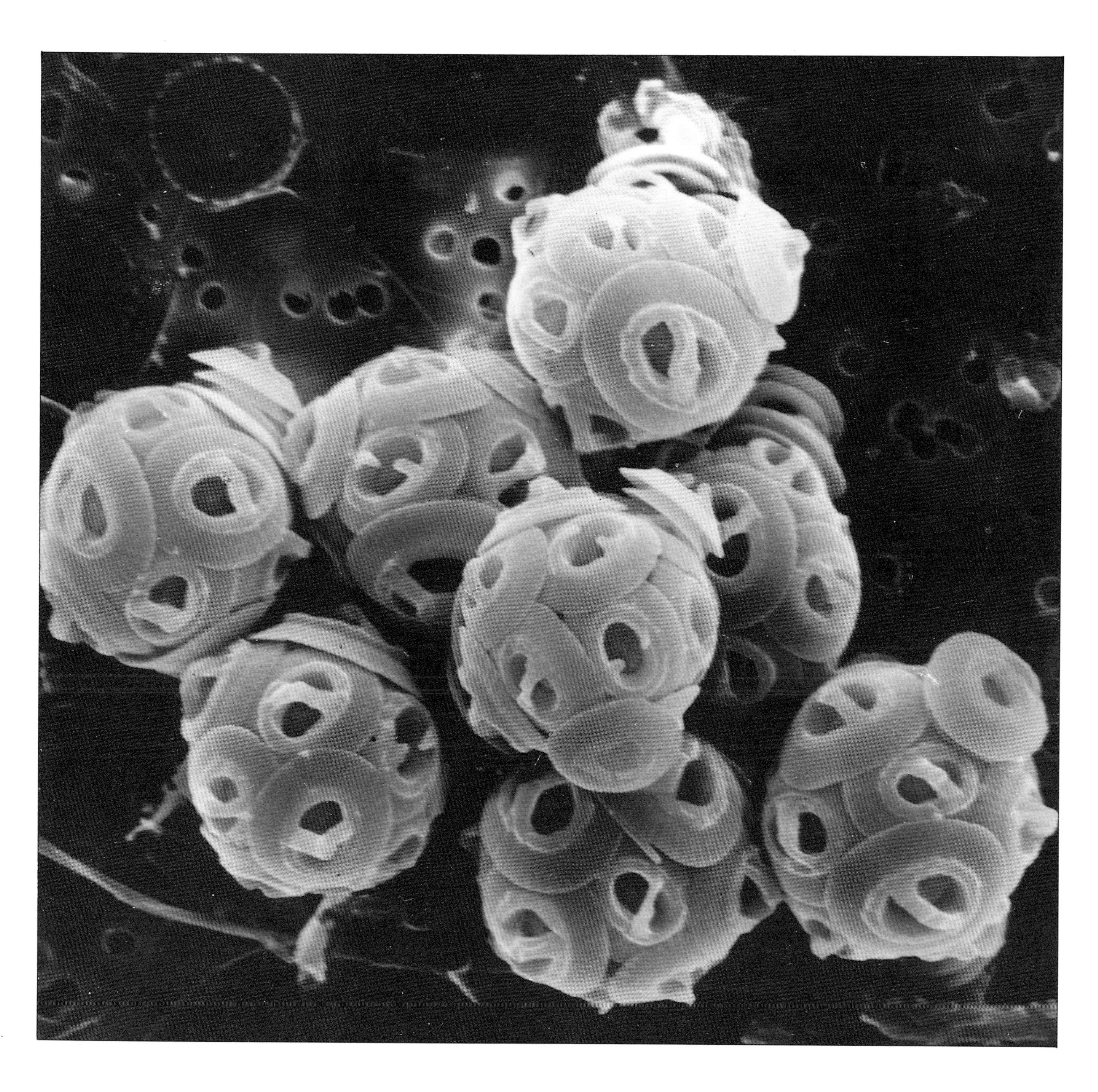

Cluster of the coccolithophorid *Gephyrocapsa oceanica*; Sydney coastal waters; individual cells 15μm in diameter.

Coccolithophorid *Emiliania huxleyi* covered by scales (coccoliths) that are composed of T-shaped elements, radially arranged as in a spoked wheel. Dense blooms of this species in European waters can result in milky-white seas; Sydney coastal waters; diameter 15μm.

Coccolithophorid *Gephyrocapsa oceanica* covered by coccoliths with two bridge-like elements arching across a central pore; Sydney coastal waters; diameter 15μm.

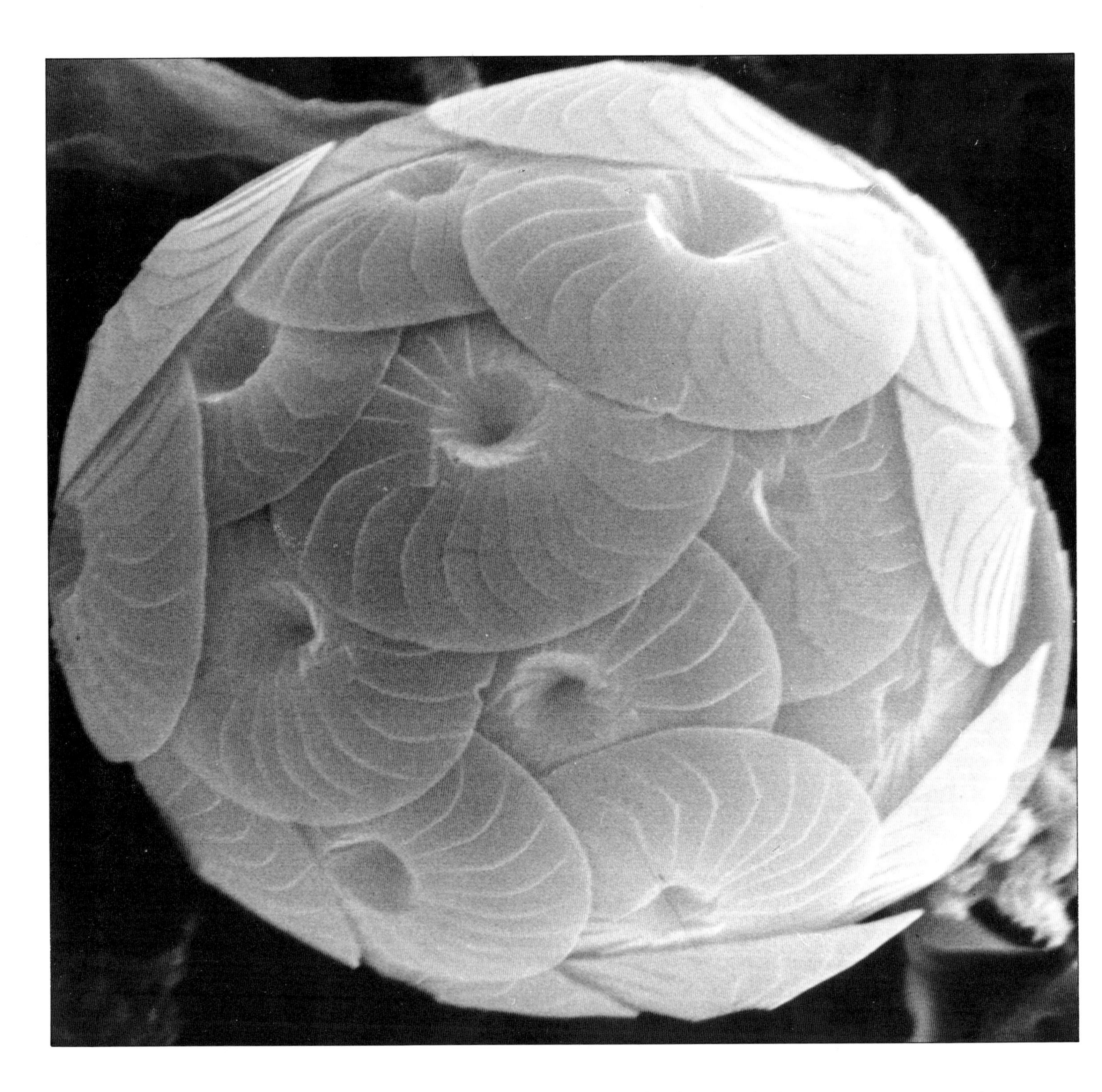

Coccolithophorid *Calcidiscus leptoporus* covered by coccoliths composed of curved lamellae; Sydney coastal waters; diameter 25μm.

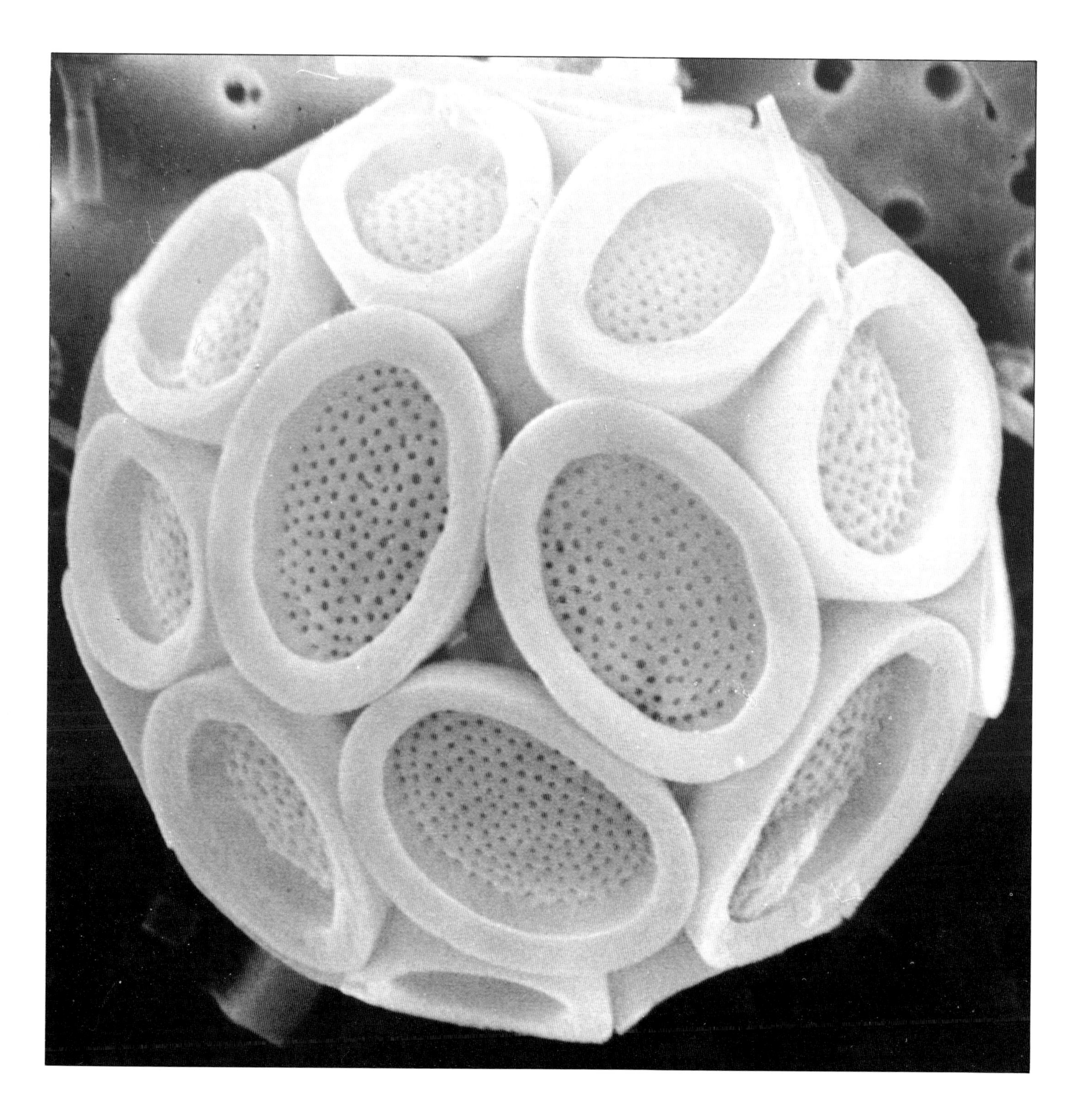

Coccolithophorid *Pontosphaera discopora* covered with strainer-shaped coccoliths; Sydney coastal waters; diameter 30μm.

Coccolithophorid *Pontosphaera japonica* covered by strainer-shaped coccoliths with broad rim; East Australian Current; diameter 20μm.

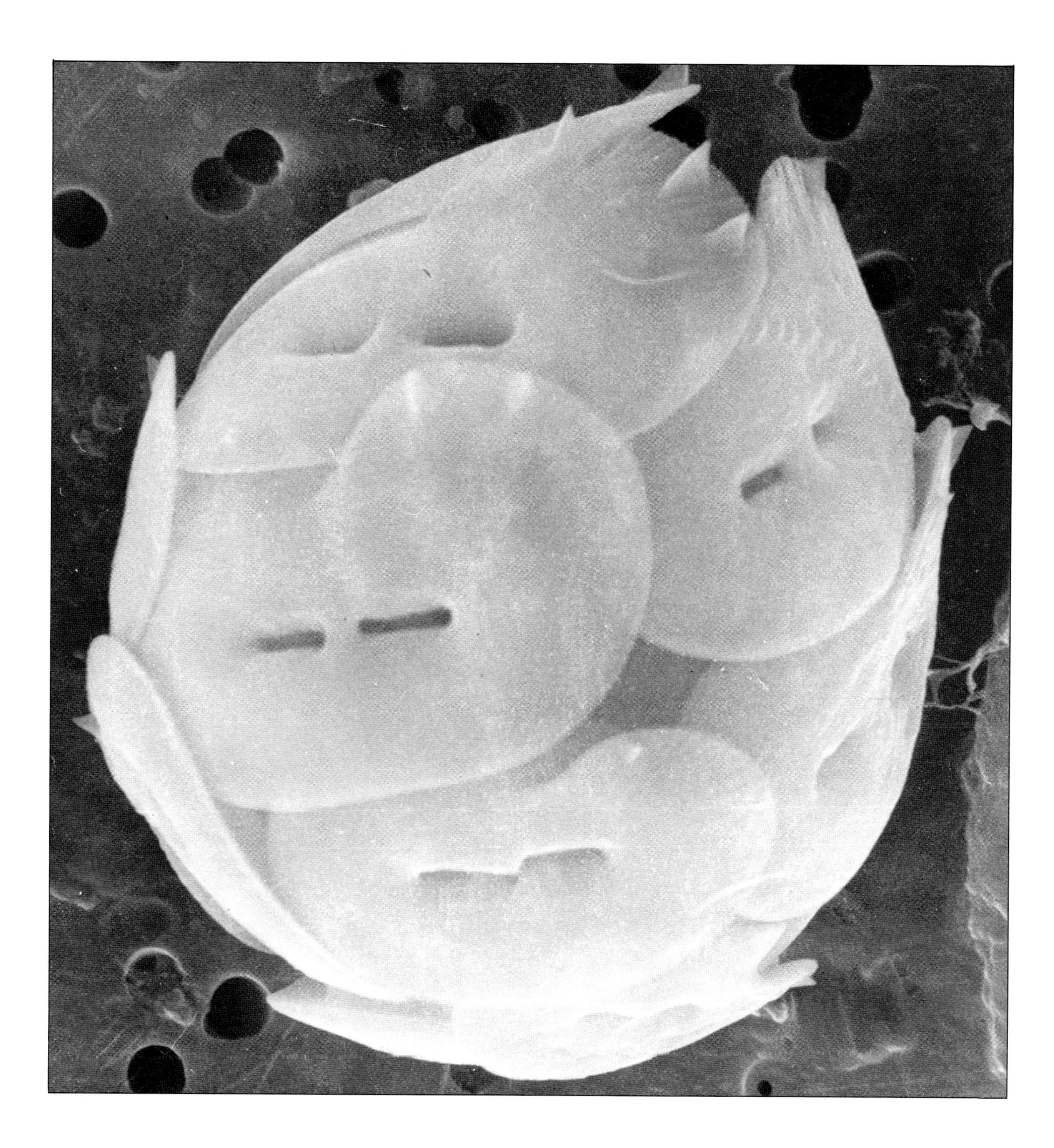

Coccolithophorid *Helicosphaera carteri* named for the spiral (helical) arrangement of coccoliths; East Australian Current; length 20μm.

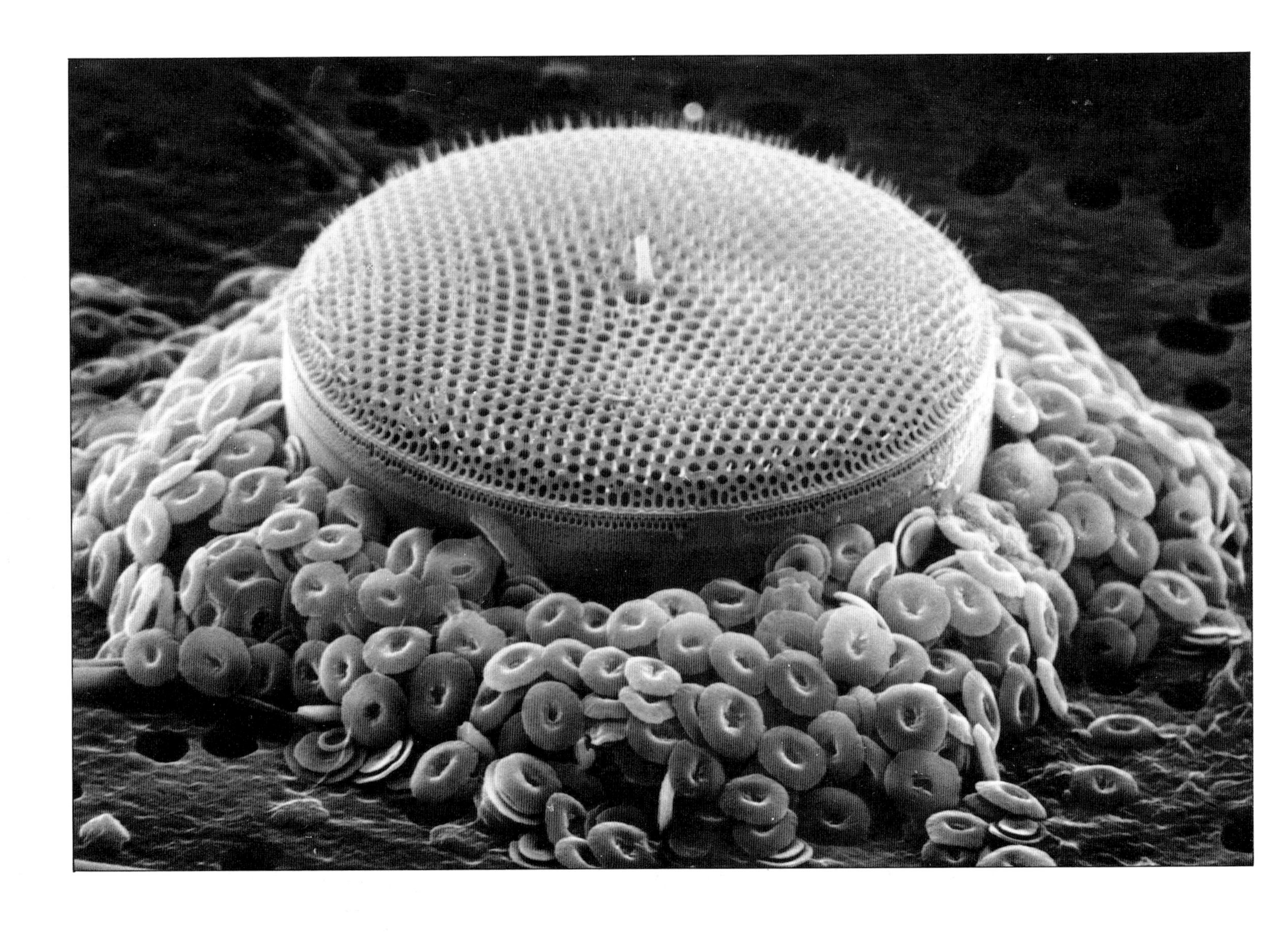

Symbiotic association (living together) of the diatom *Thalassiosira punctifera,* surrounded with a crown of the coccolithophorid *Crenalithus sessilis*; Coral Sea; diameter of diatom 70μm.

Bottom-dwelling diatom *Actinoptychus senarius* (Actino = ray; ptychus = fold); North West Shelf of Australia; diameter 30μm.

Diatom *Asteromphalus elegans* with spider-like pattern of rays (Astero = star; omphalus = navel); North West Shelf of Australia; diameter 40μm.

Inside view of a cell of the diatom *Asteromphalus heptactis*; showing hollow flotation tubes, Coral Sea; diameter 30μm.

Warm-water diatom *Planktoniella sol* (Plankto = wandering; sol = sun), similar to *Thalassiosira eccentrica,* except for the marginal wing which is thought to act as a flotation device; North West Shelf of Australia; diameter 60 μm.

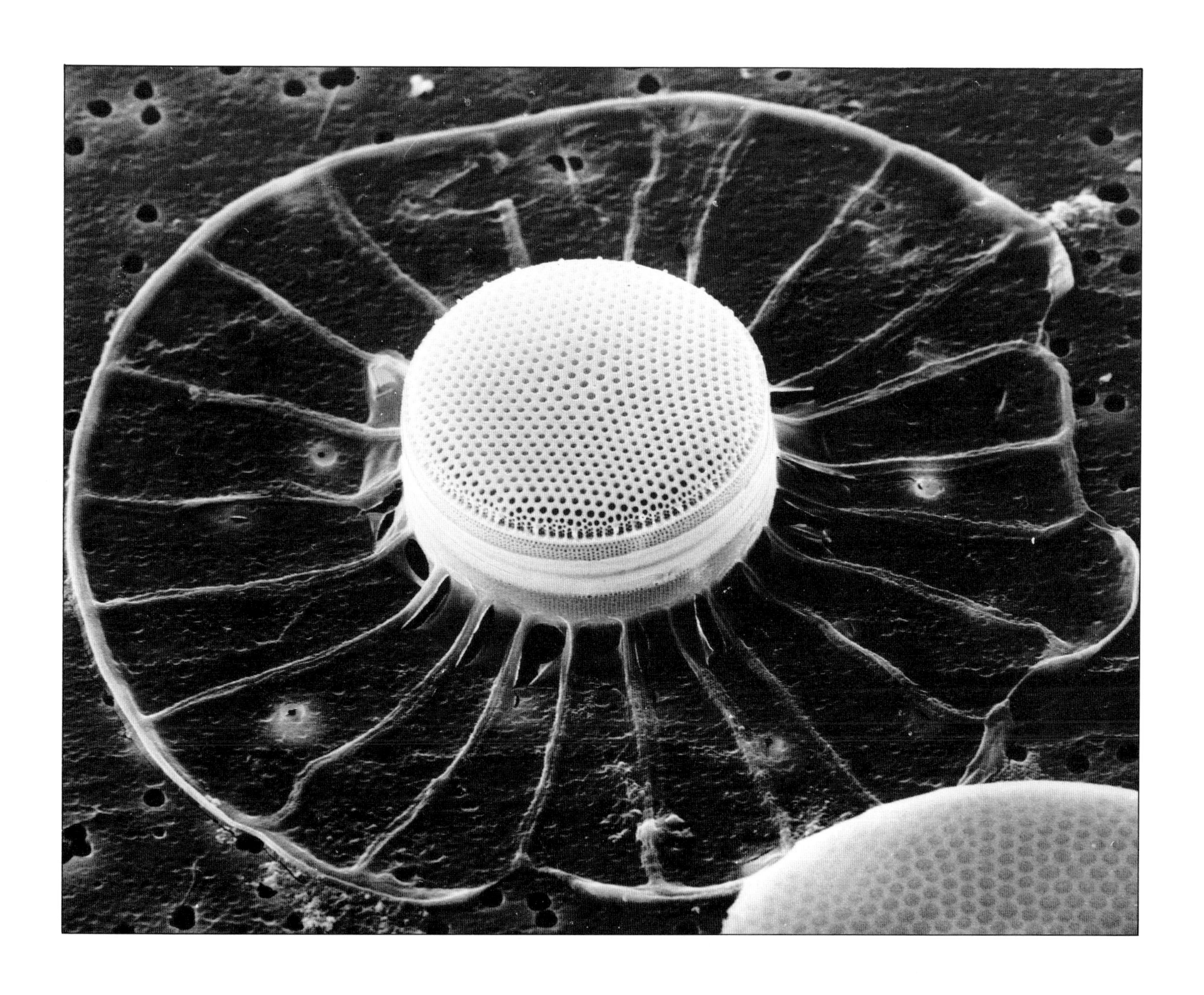

Diatom *Thalassiosira eccentrica* (Thalassio = sea; sira = thread) with eccentric pattern of pores (areolae) and marginal ring of spines; North West Shelf of Australia; diameter 40μm.

Bottom-dwelling diatom *Auliscus sculptus* (Auliscus = groove) with two eye-like ocelli (circular hole) and grooved sculpturing; Spring Bay, Tasmania; diameter 40 μm.

Pill-box shaped diatom *Thalassiosira oestrupii* (oestrupii = dedicated to E. Oestrup), the top cell in lateral (girdle) view and the bottom cell in frontal (valve) view; North West Shelf of Australia; diameter 25 μm.

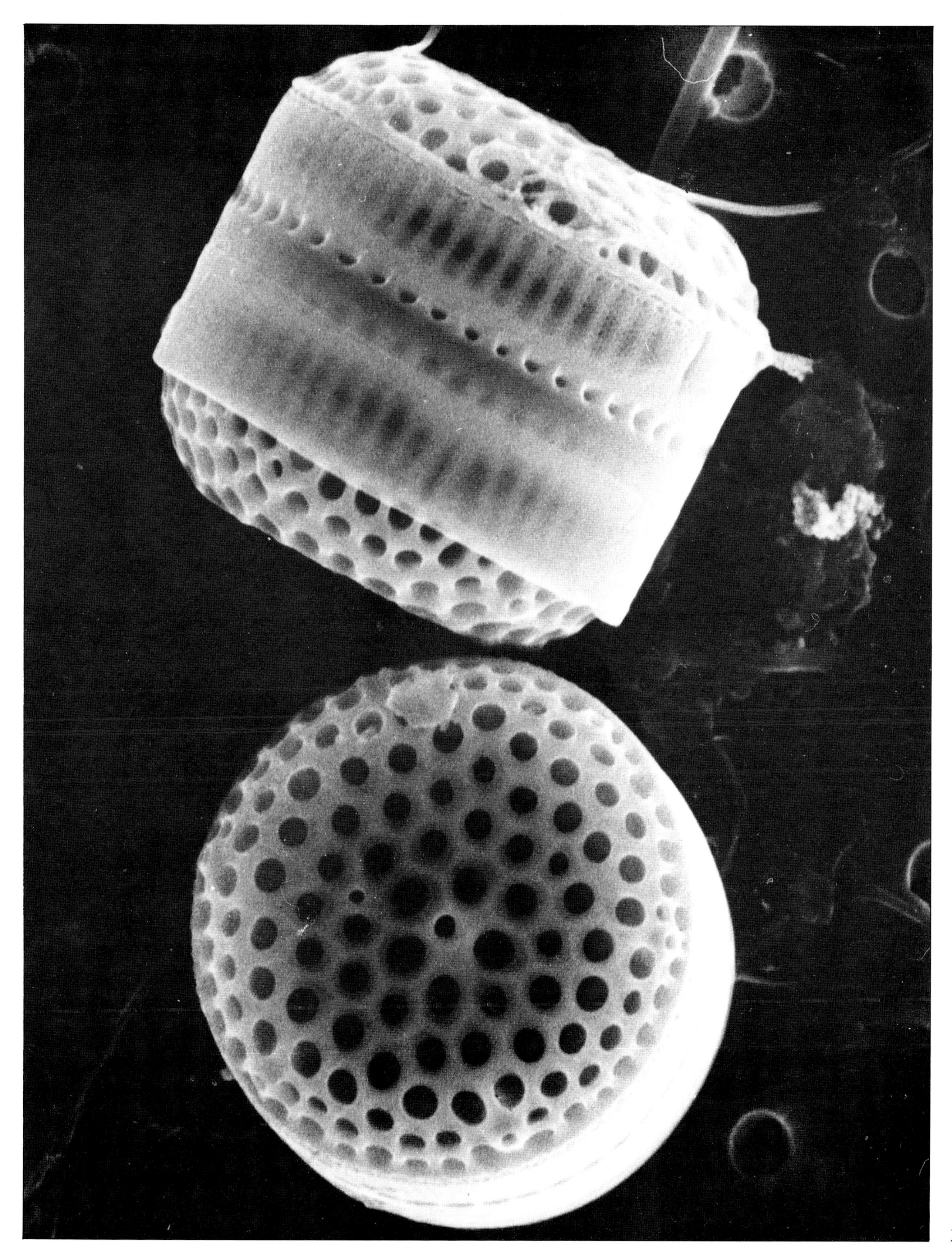

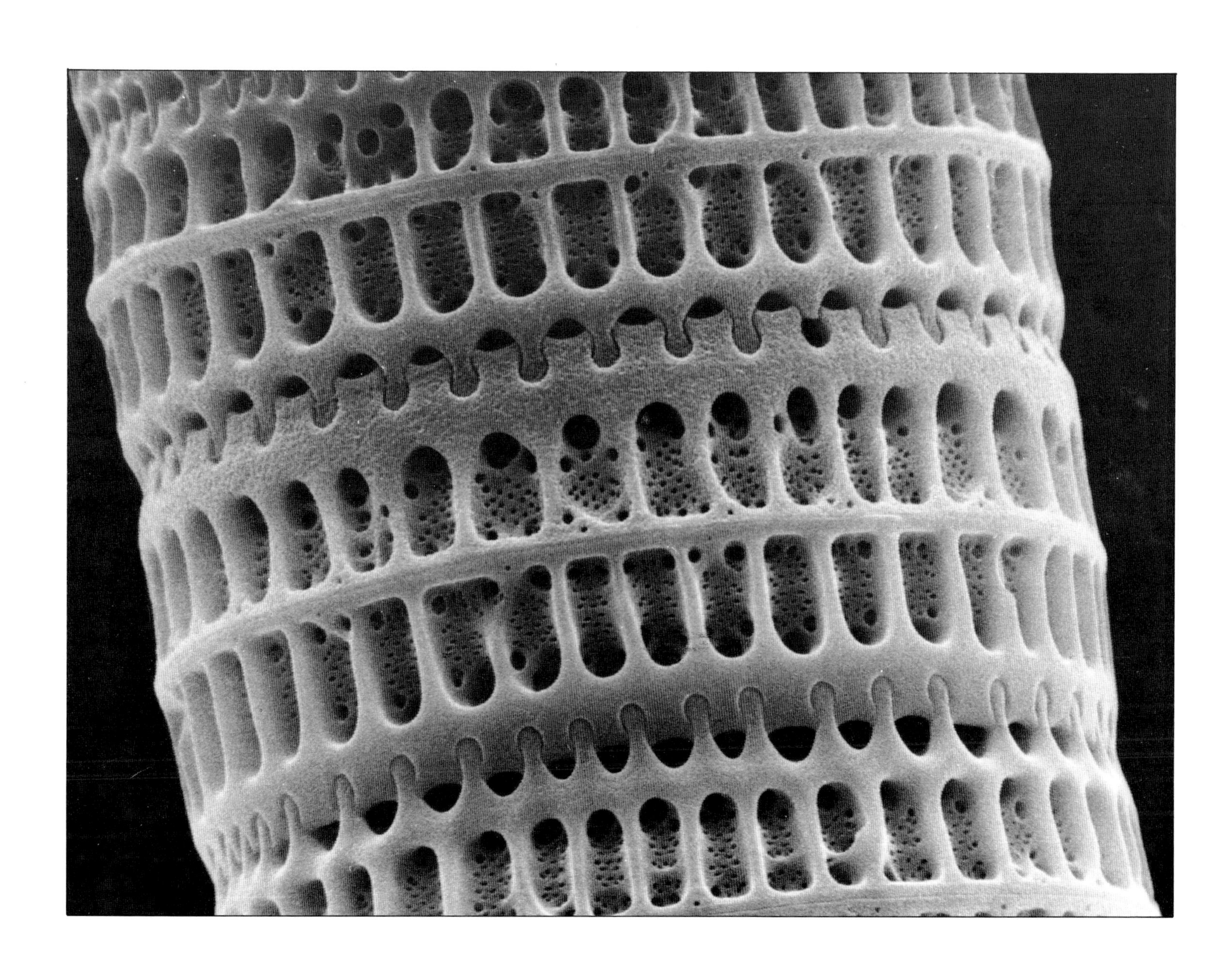

Diatom *Thalassiosira allenii* (allenii = named in honour of E.J. Allen), the top cell in valve view and the bottom cell in girdle view. Mucus threads exuded from the central and marginal (strutted) processes are used for colony formation; Sydney coastal waters; diameter 15 μm.

Bottom-dwelling diatom *Paralia sulcata* (Paralia = shore) in girdle view showing interlocking cells of a chain; Huon River, Tasmania; diameter 30 μm.

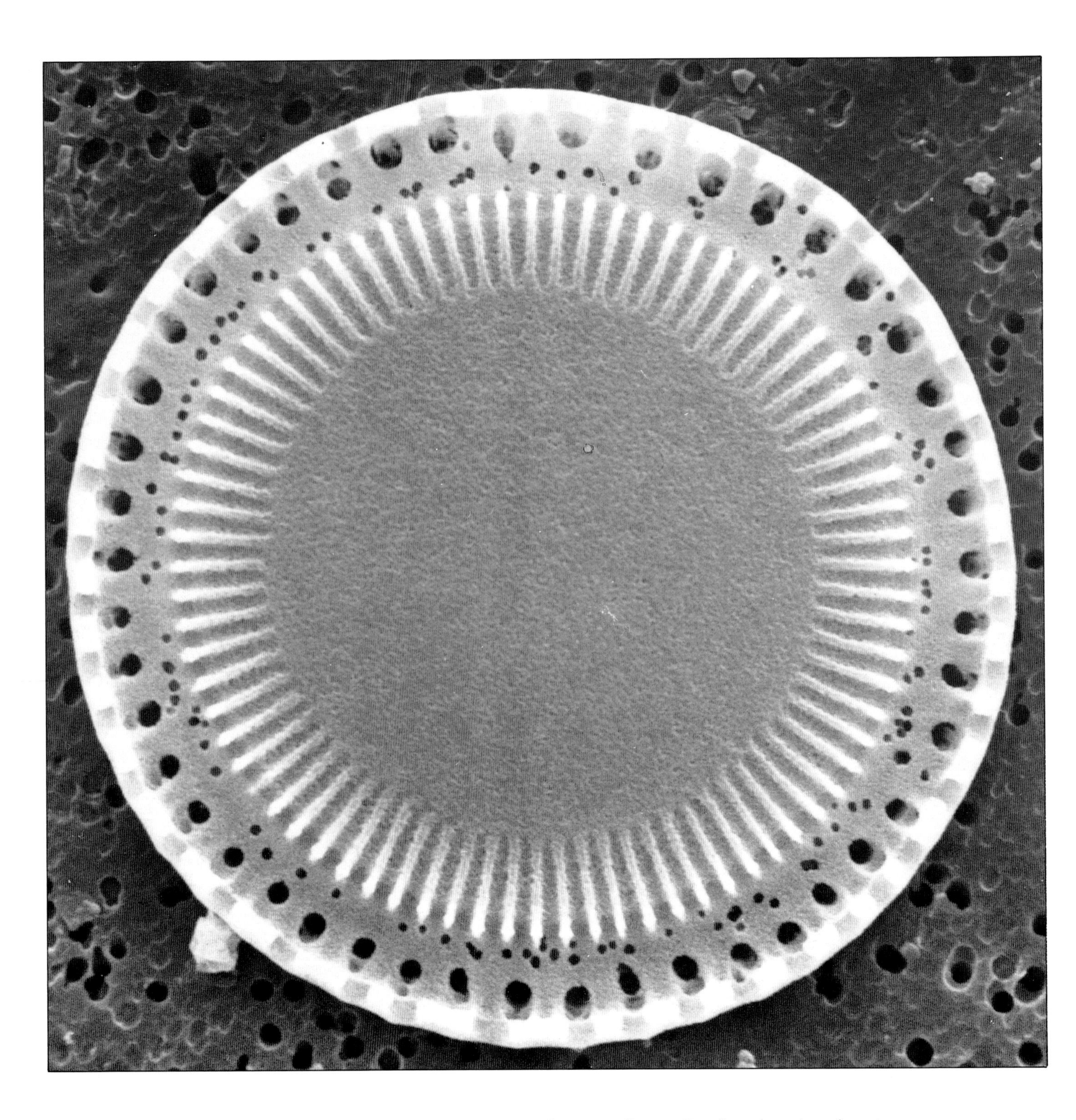

The same diatom *Paralia sulcata* in valve view.

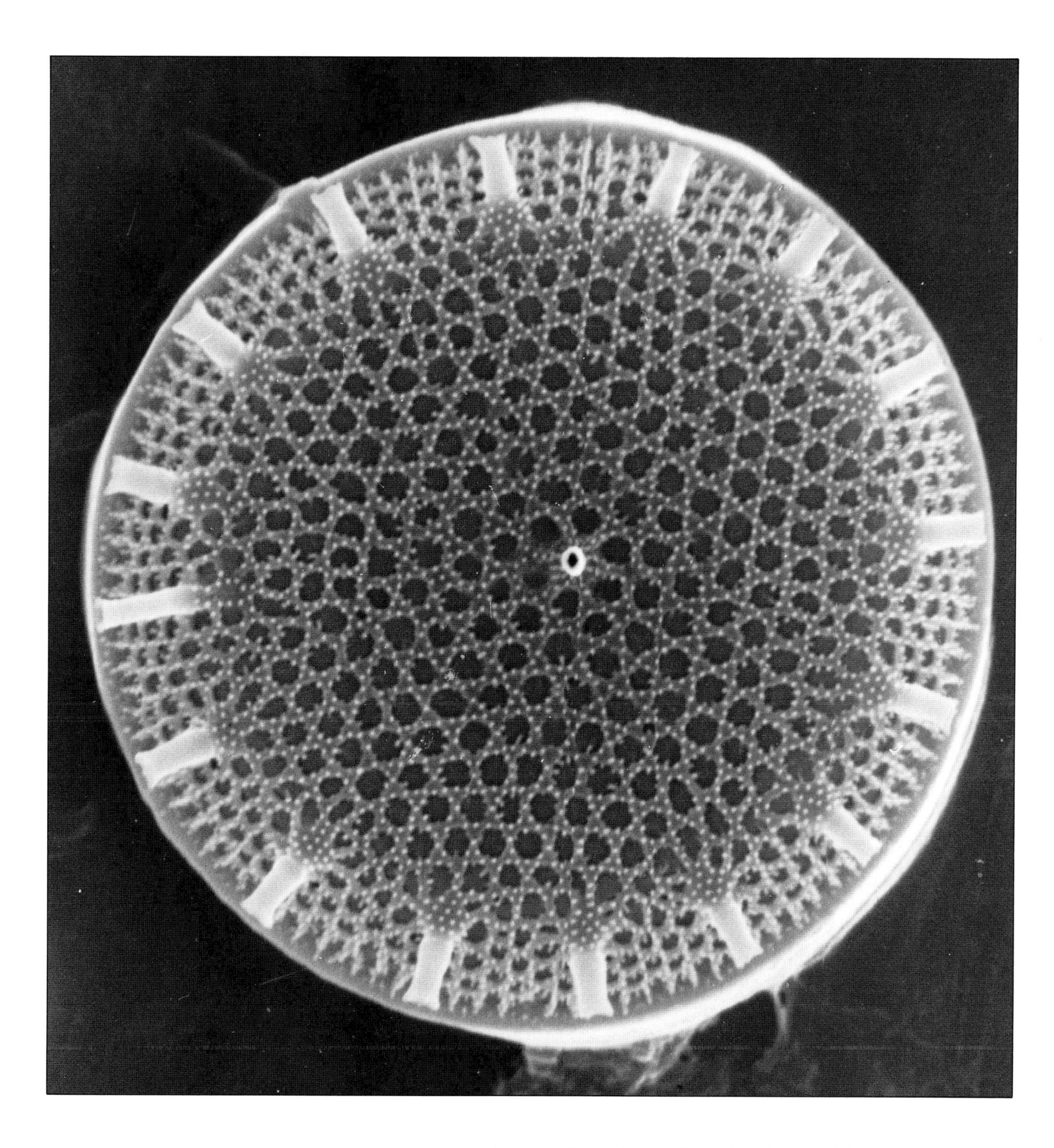

Diatom *Thalassiosira allenii* showing lace-like ornamentation of the silica cell wall; Port Adelaide; diameter 15 μm.

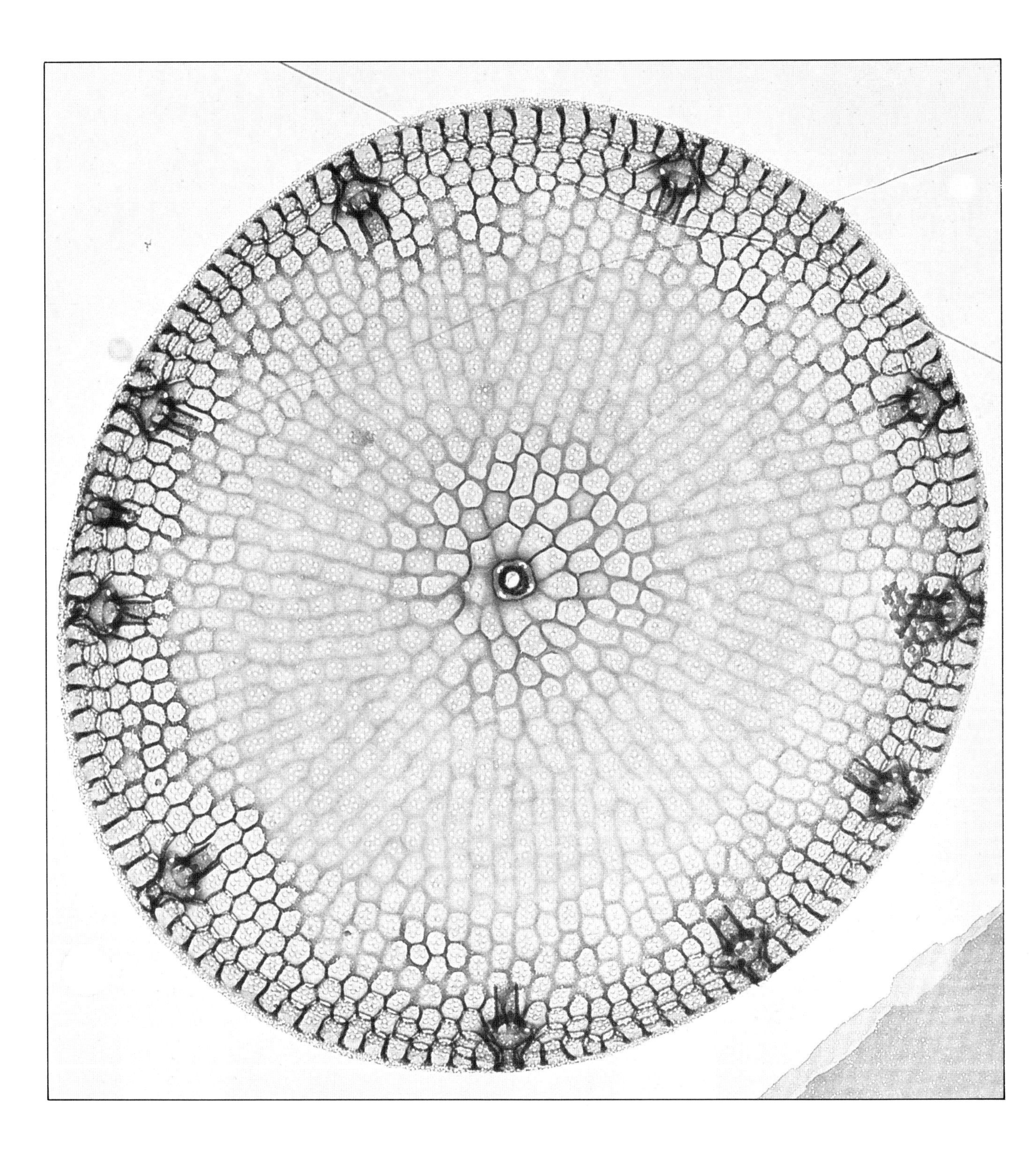

Thinly silicified cell of the diatom *Thalassiosira partheneia* seen in the transmission electron microscope; Gulf of Carpentaria; diameter 10 μm.

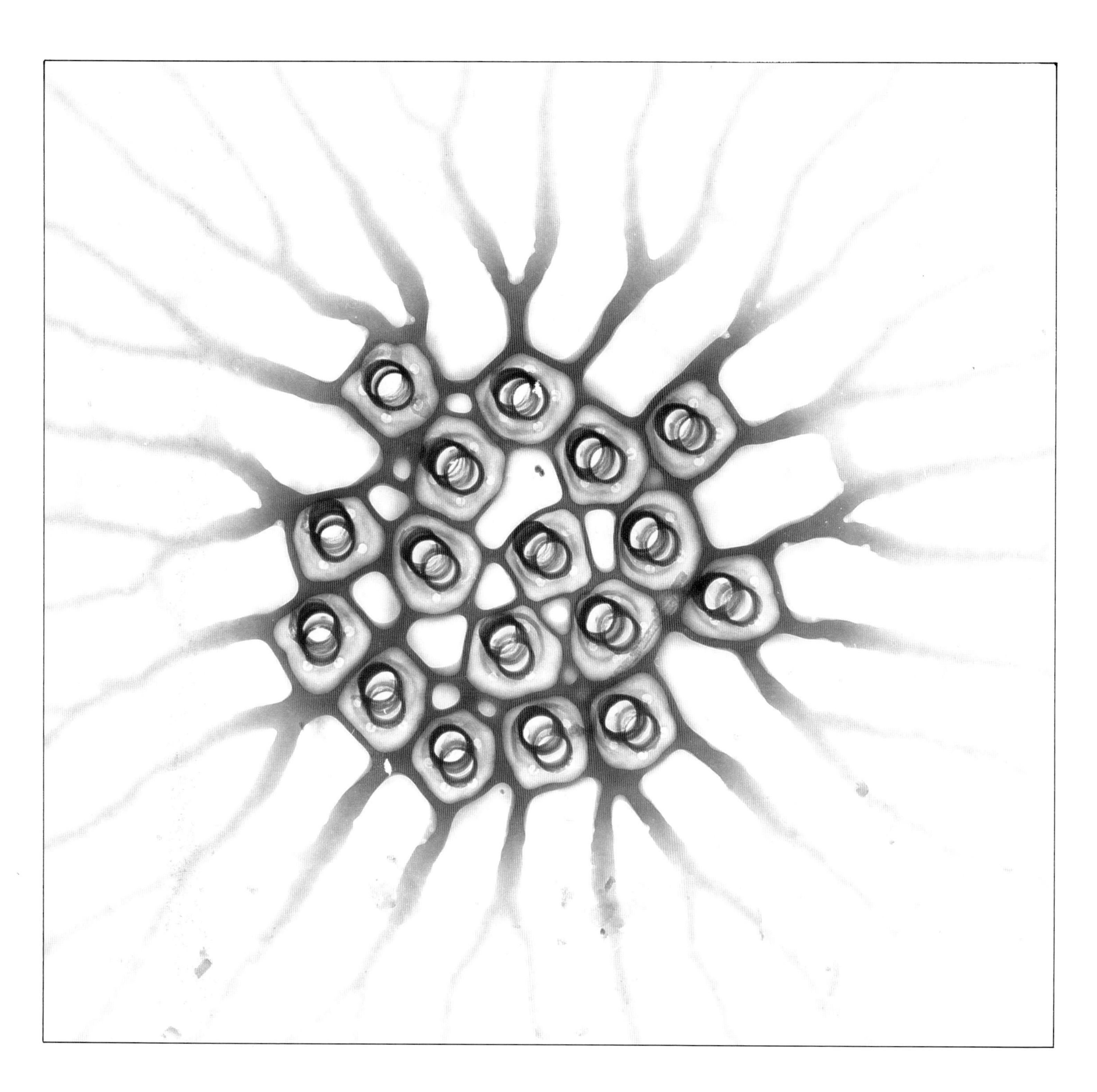

Detail of central tubular (strutted) processes of the diatom *Thalassiosira rotula*. These tubes exude mucus threads used for chain formation; Sydney coastal waters.

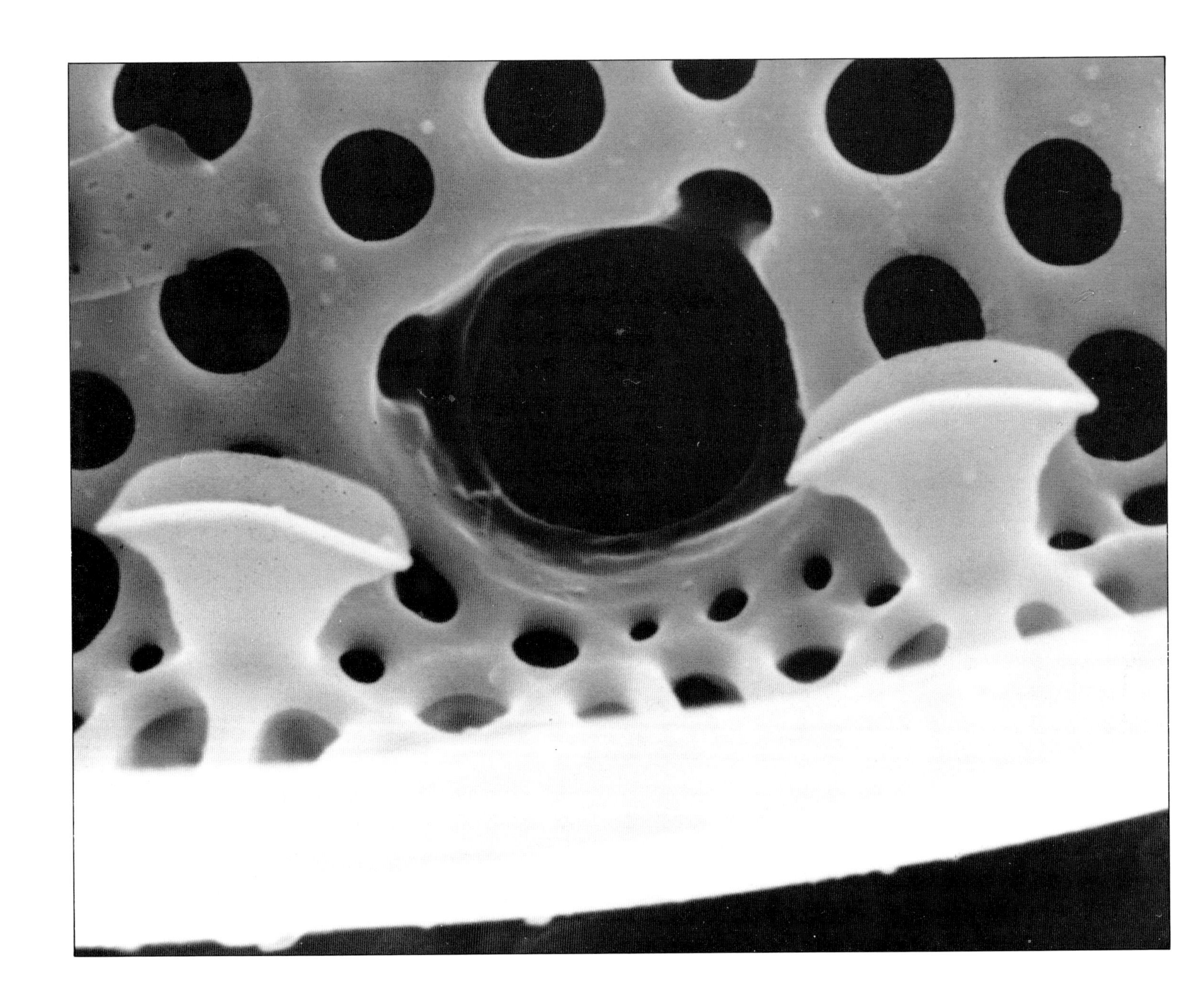

Detail of cell margin of the diatom *Roperia tesselata* (Roperia = dedicated to F.C.S. Roper) showing two lip-shaped (labiate) processes. next to a circular hole (pseudonodulus); Sydney coastal waters.

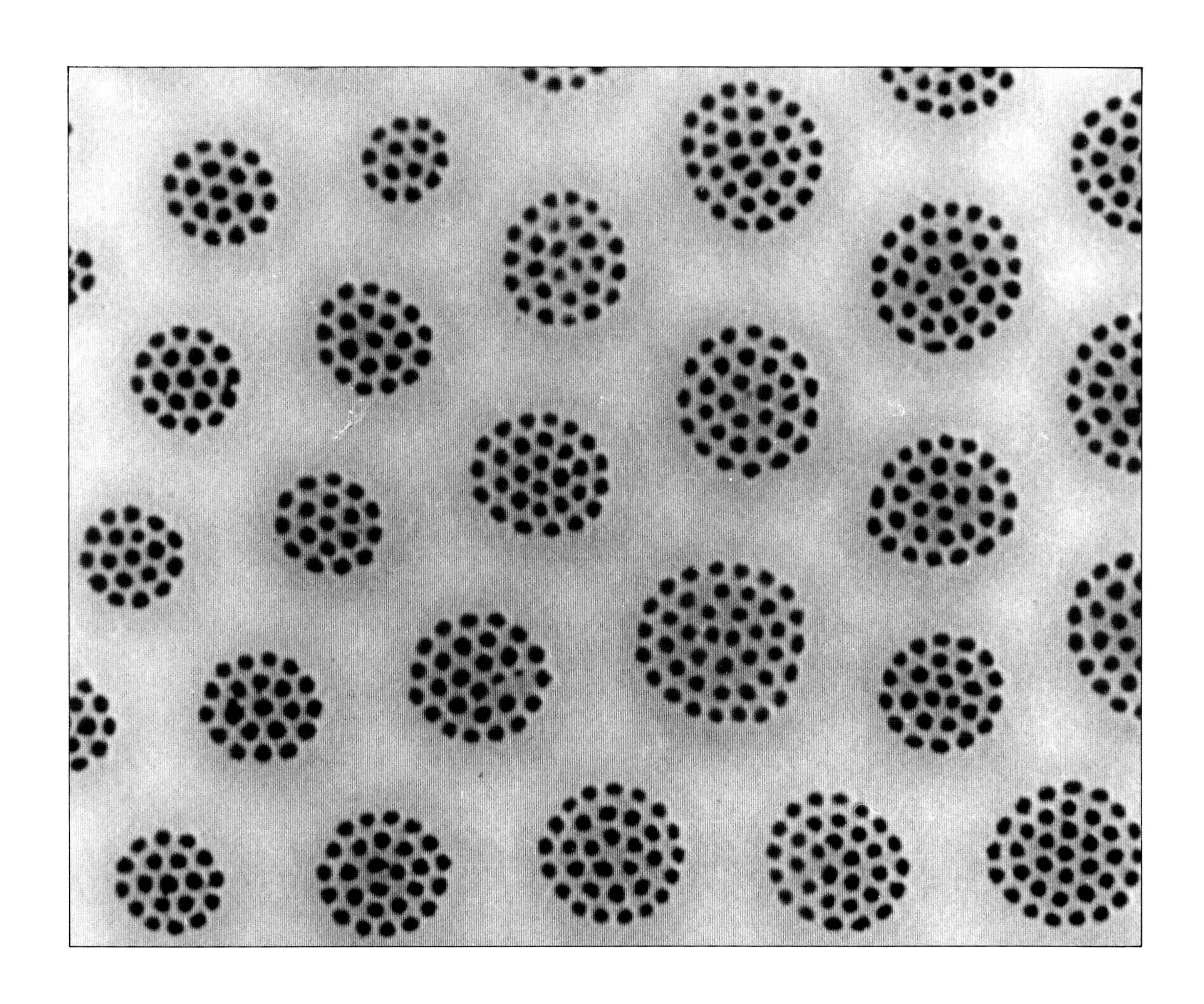

Detail of ornamentation of the diatom *Coscinodiscus radiatus* (Coscino = sieve; discus = disc). Delicate pores in the silica cell wall allow contact between cell contents and the surrounding sea water; North West Shelf of Australia.

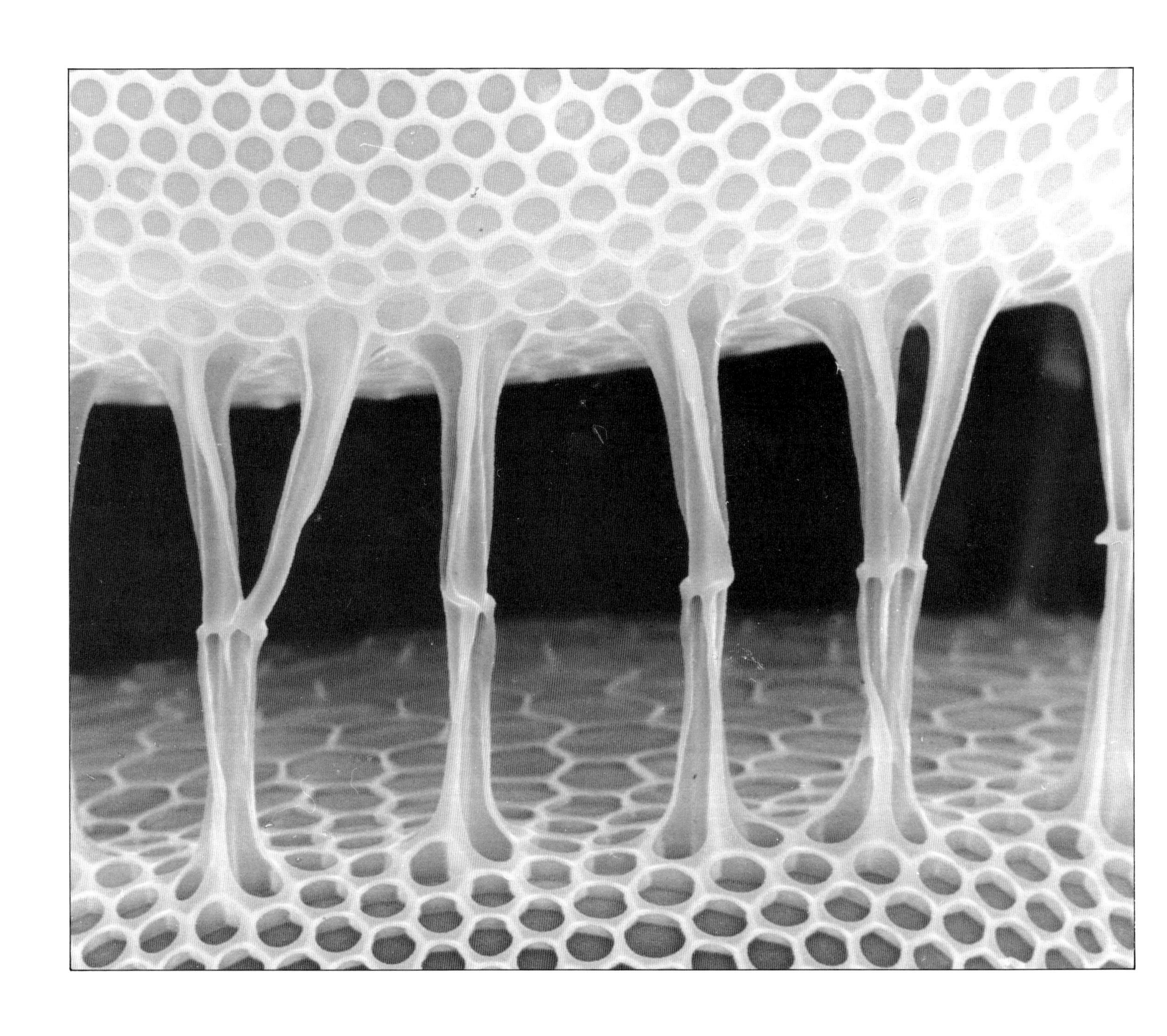

Pillar-like connection between two cells of the chain-forming diatom *Stephanopyxis palmeriana* (Stephanos = crown; pyxis = box); North West Shelf of Australia.

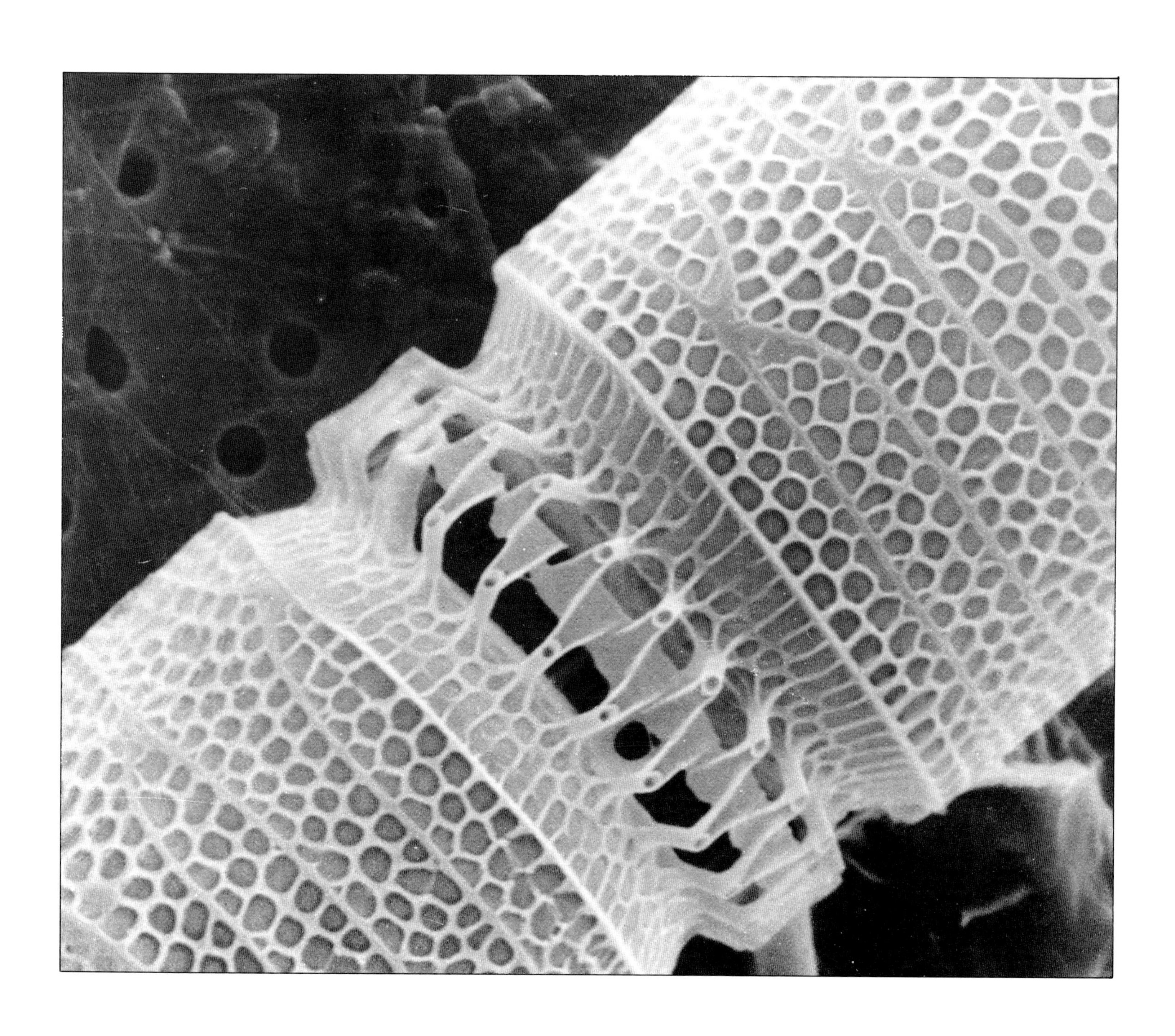

Lace-like interlocking connection between two cells of the chain-forming diatom *Detonula pumila* (Detonula = dedicated to G.B. De Toni); Sydney coastal waters.

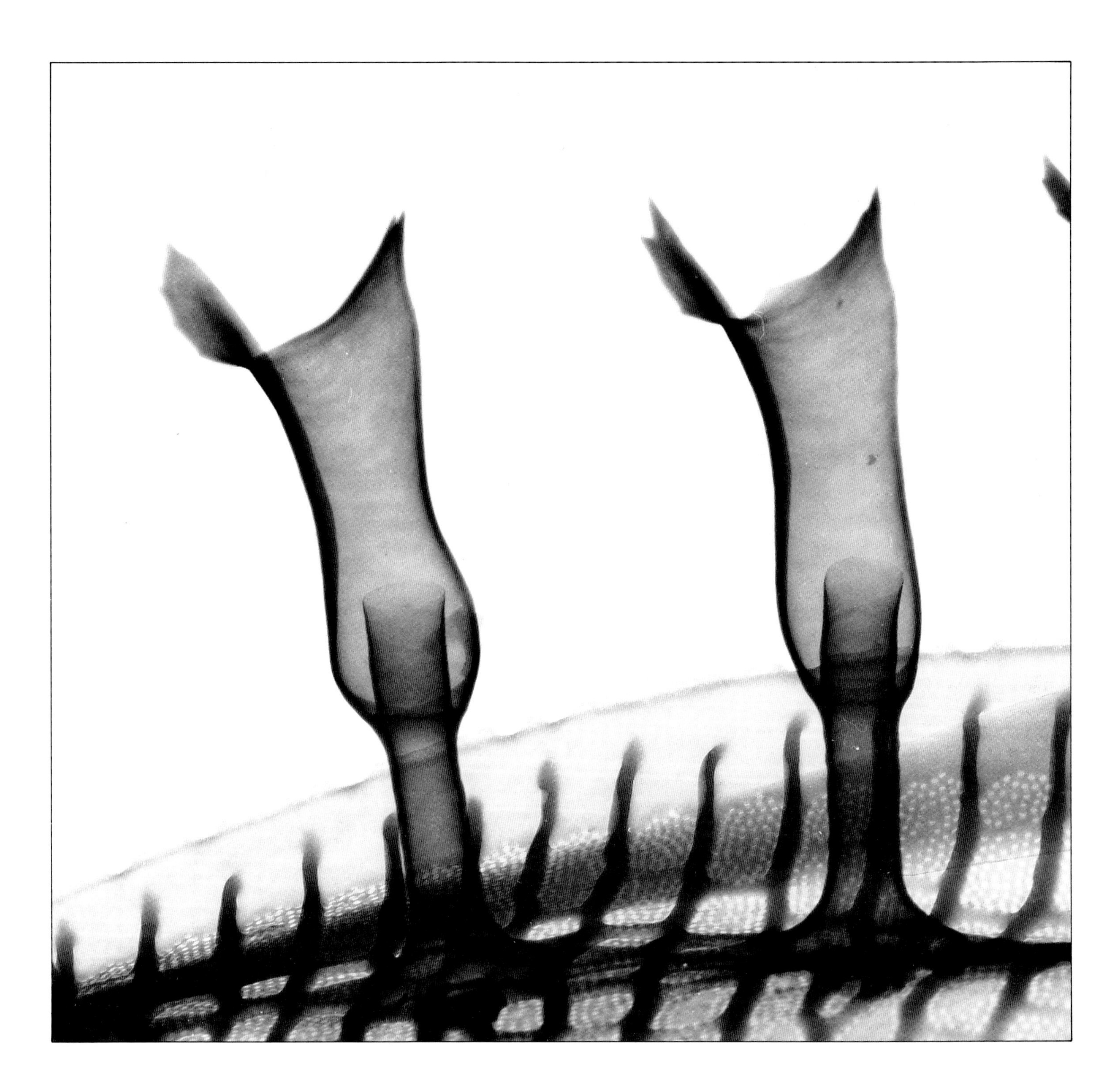

Detail of tulip-shaped, tubular (strutted) processes of the diatom *Detonula pumila* seen in the transmission electron microscope; Sydney coastal waters.

Detail of tubular (occluded) processes that connect two cells of the chain forming diatom *Lauderia annulata* (Lauderia = dedicated to H.Q. Lauder); Sydney coastal waters.

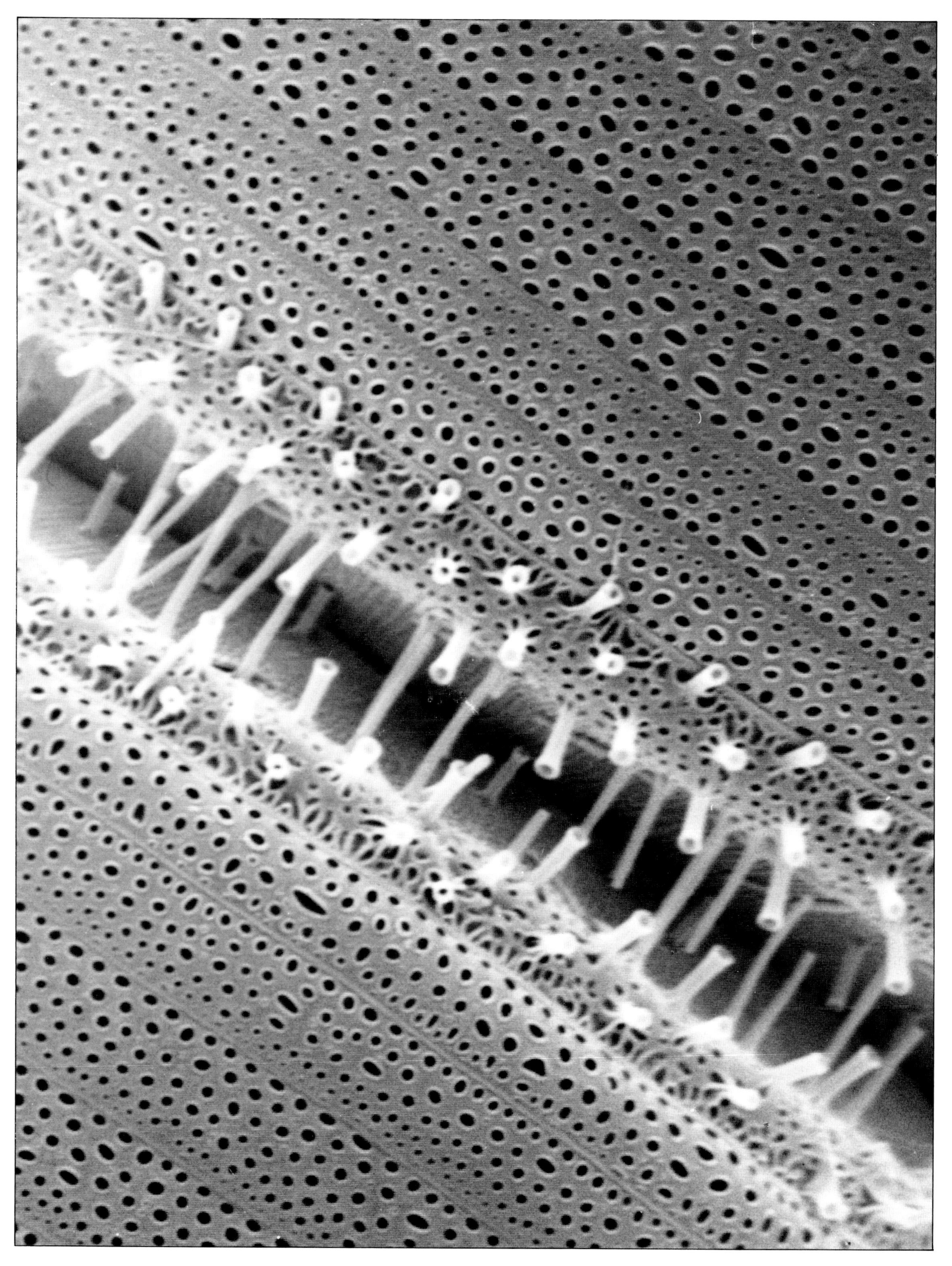

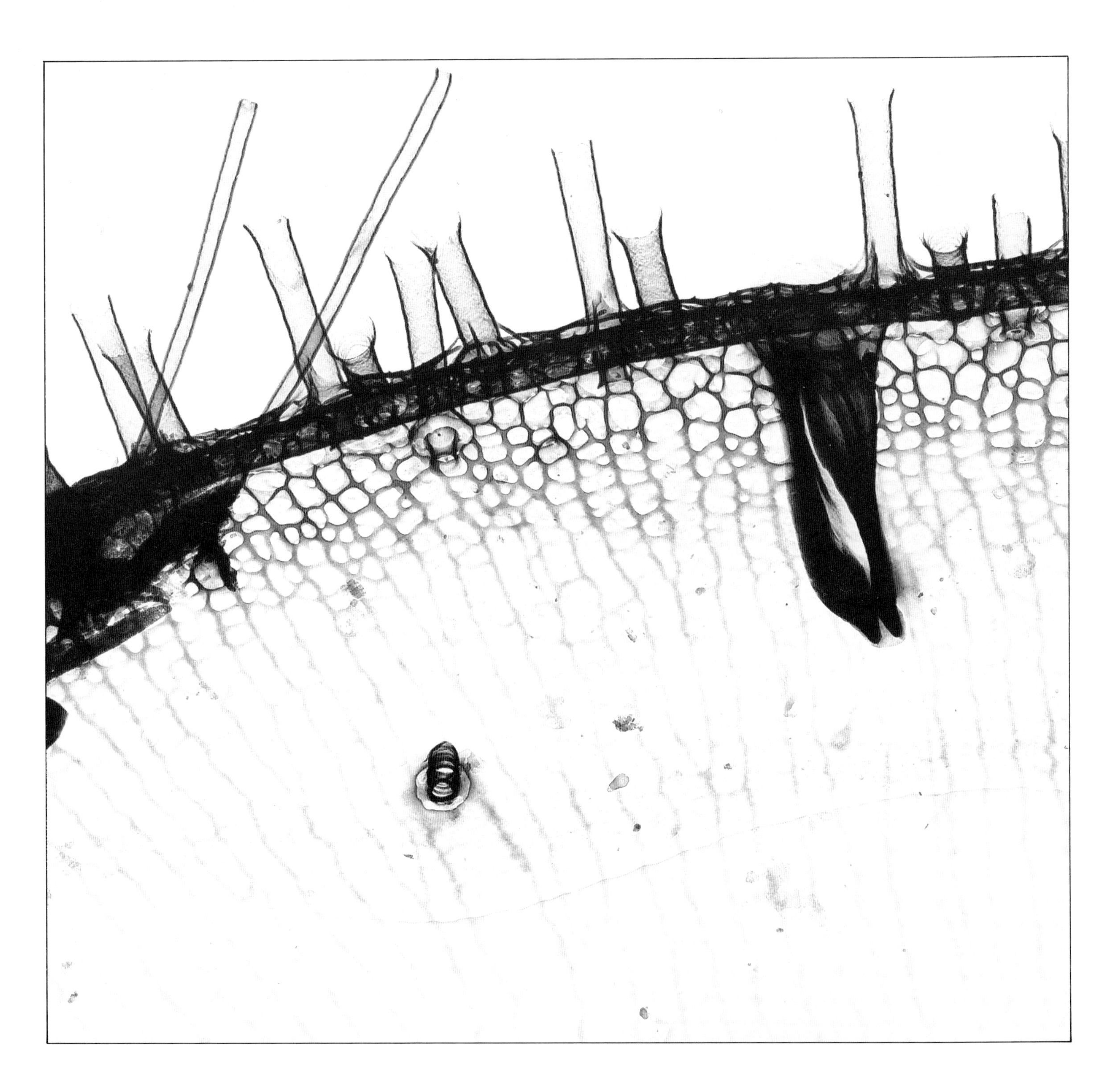

Detail of the occluded processes of *Lauderia annulata* seen in the transmission electron microscope; Sydney coastal waters.

Chain-forming diatom *Leptocylindrus mediterraneus* (Lepto = narrow; cylindrus = cylinder) showing zig-zag configuration of girdle bands; North West Shelf of Australia; diameter 30 μm.

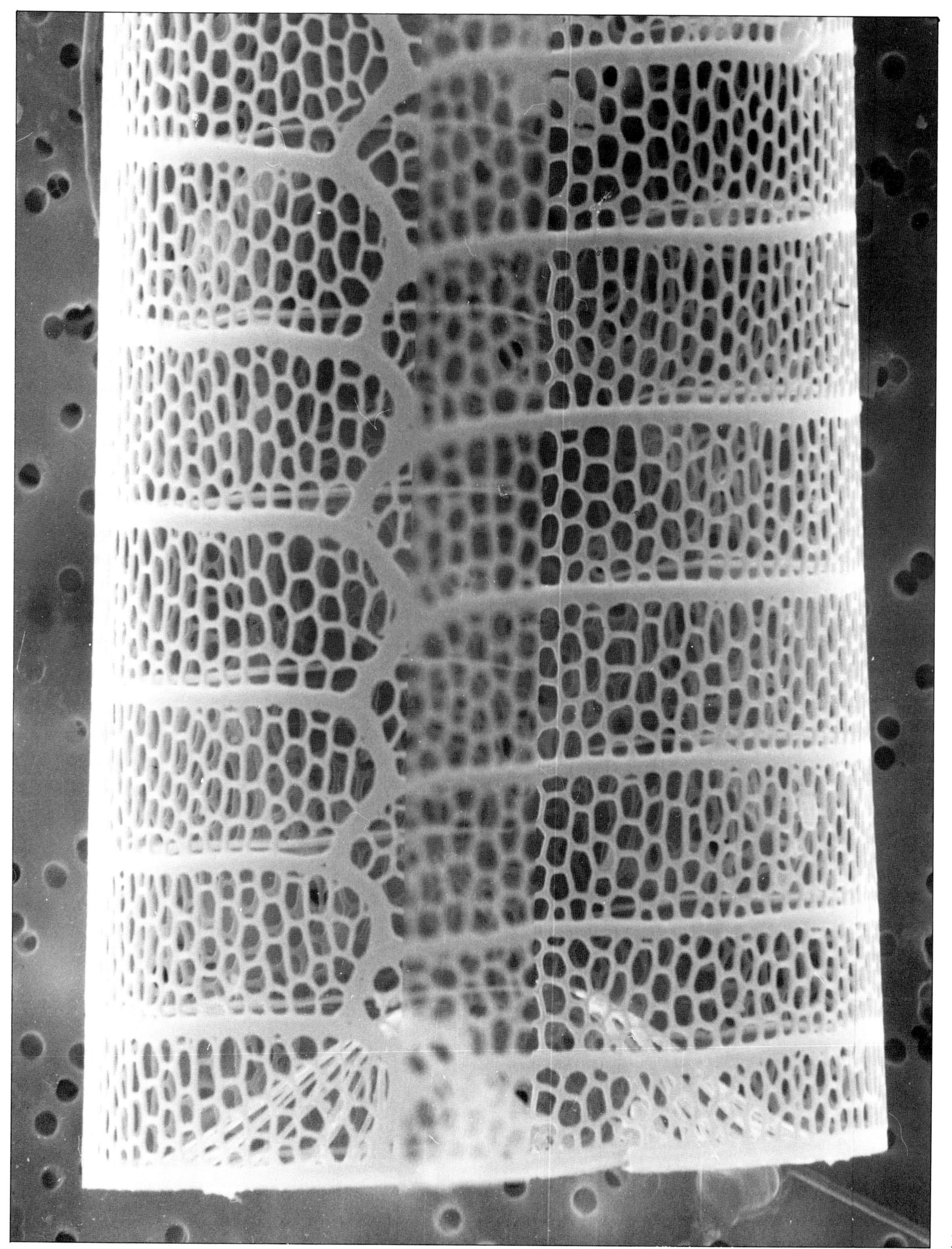

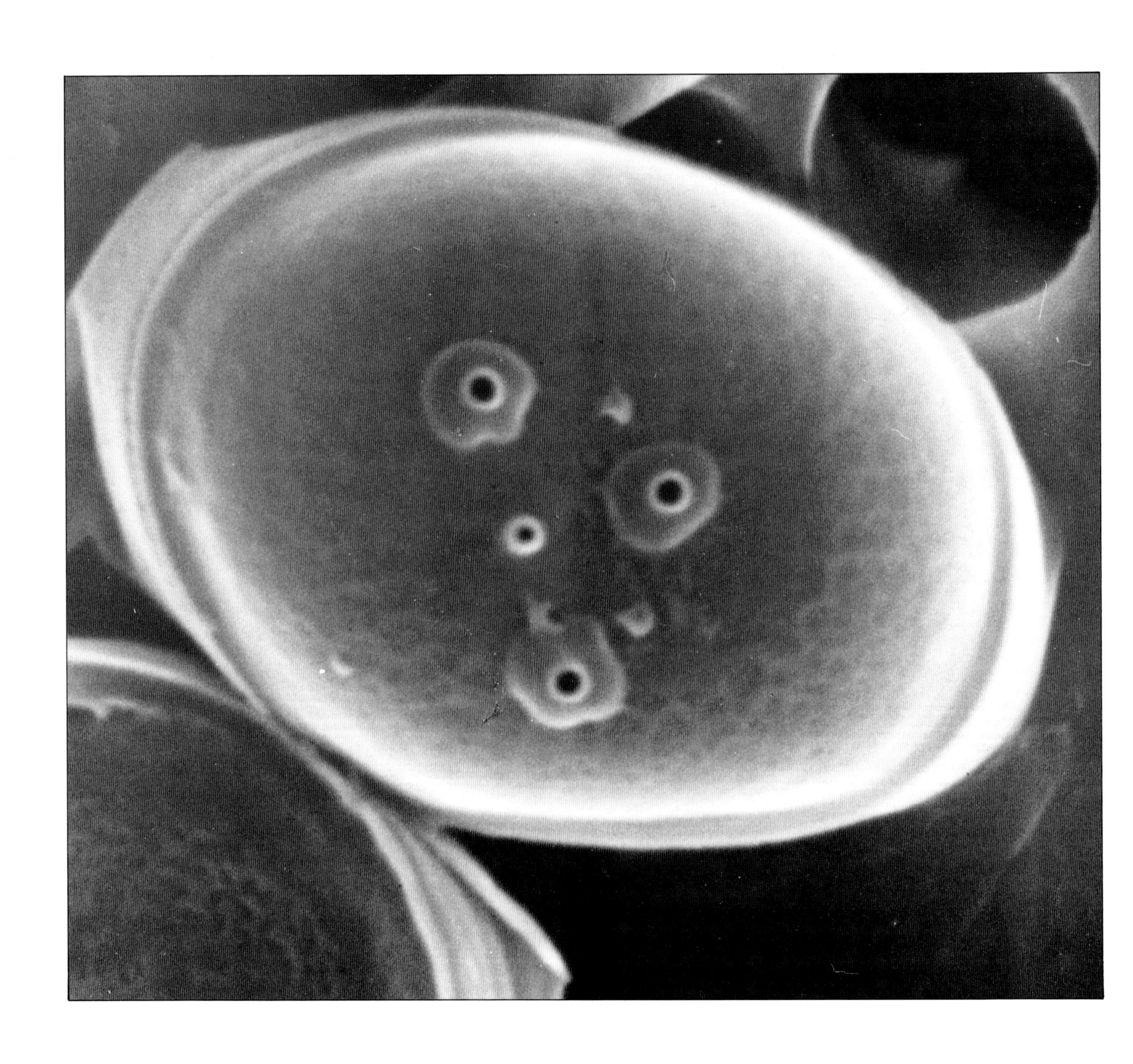

Diatom *Minidiscus trioculatus* (Mini = small; discus = disc; trioculatus = three eyes) next to a 1 micron pore (top left) in a millipore filter used to collect this minute plankton organism; Sydney coastal waters; diameter 3 μm.

Tropical chain-forming diatom *Bacteriastrum hyalinum* (Bacterio = small stick; astrum = star) with hair-like extensions that are thought to function as a flotation device in less dense warmer waters; North West Shelf of Australia; diameter 30μm.

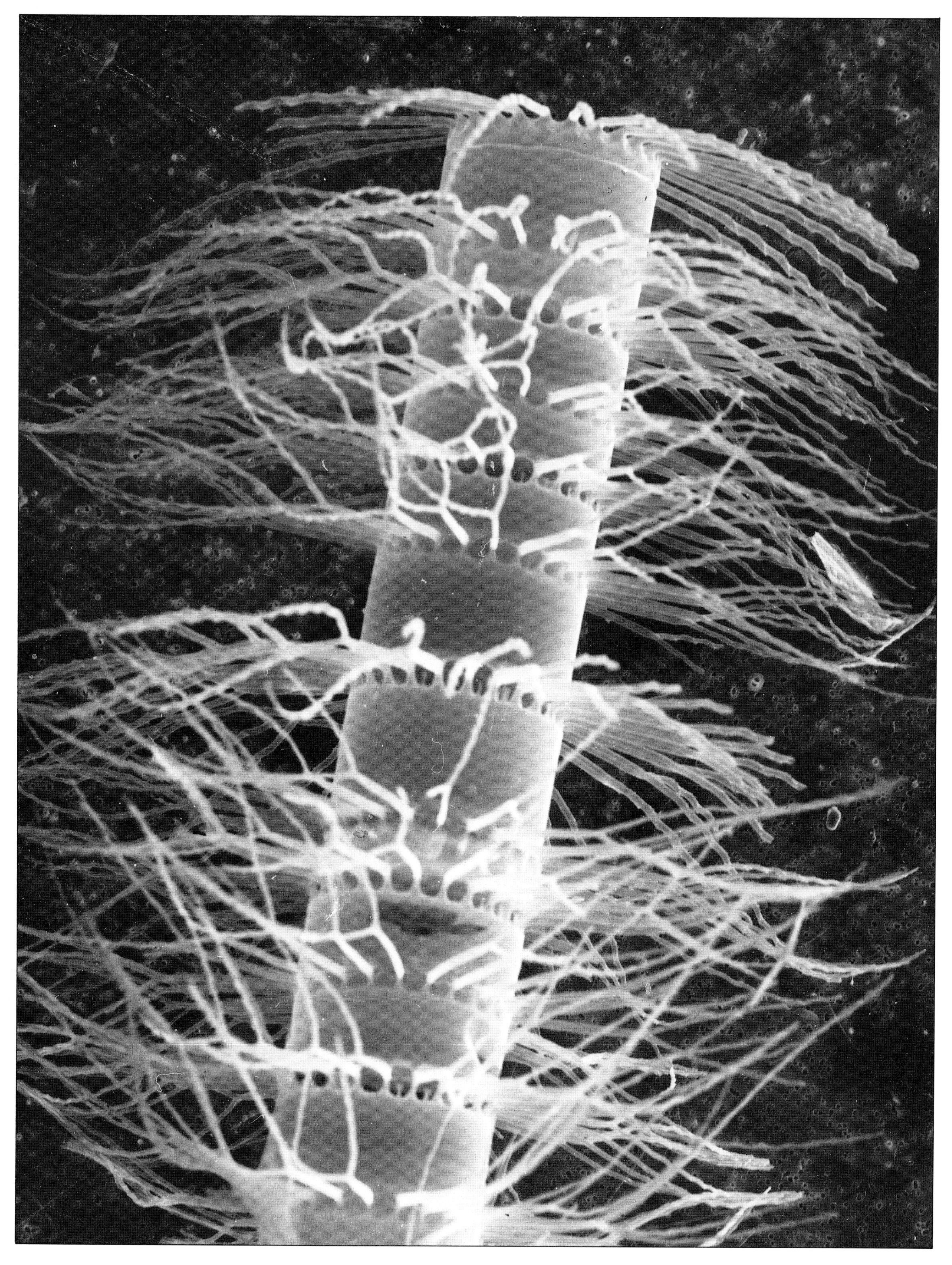

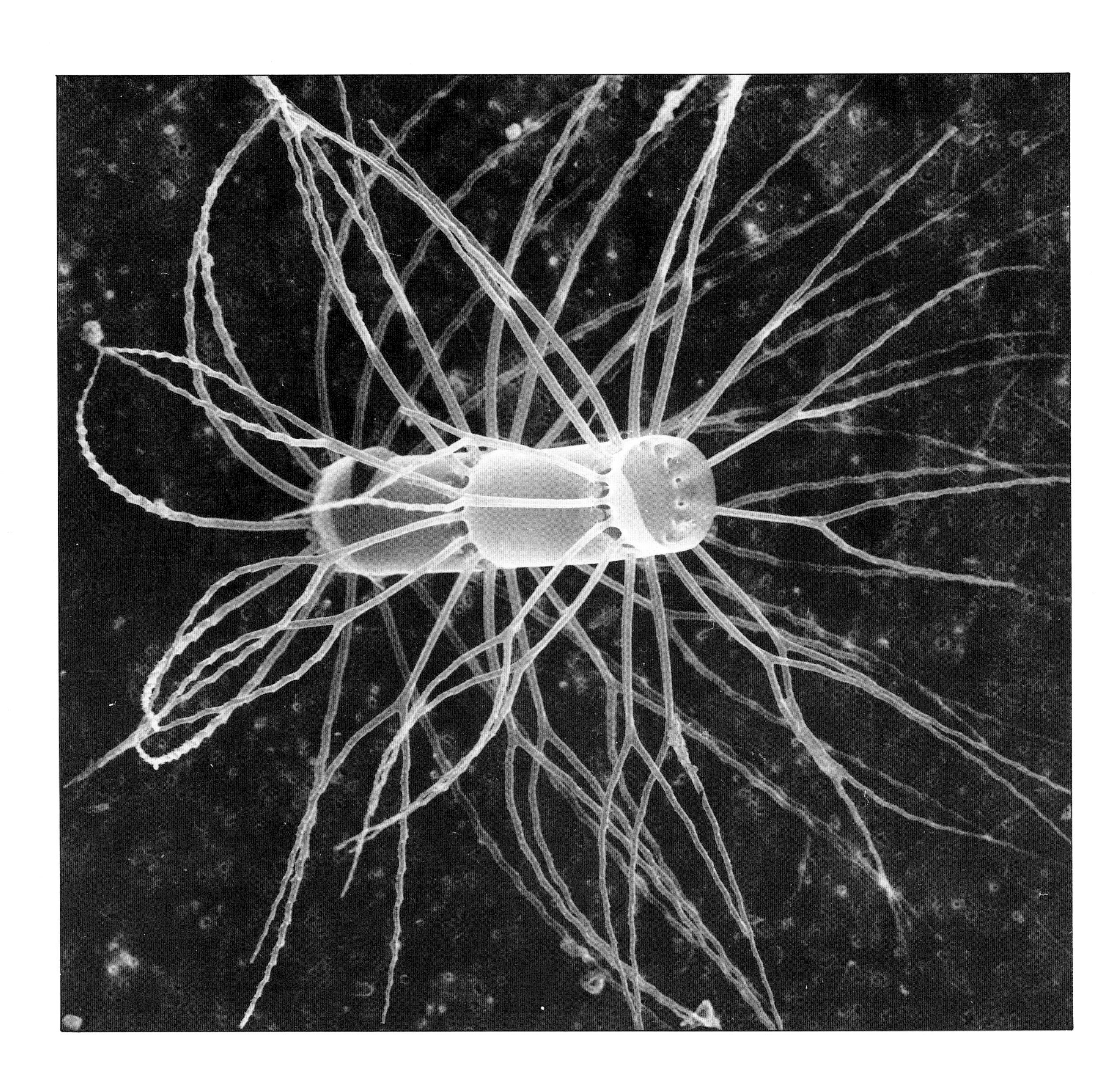

Tropical chain-forming diatom *Bacteriastrum furcatum* named for the forked hair-like extensions; North West Shelf of Australia; diameter 30 μm.

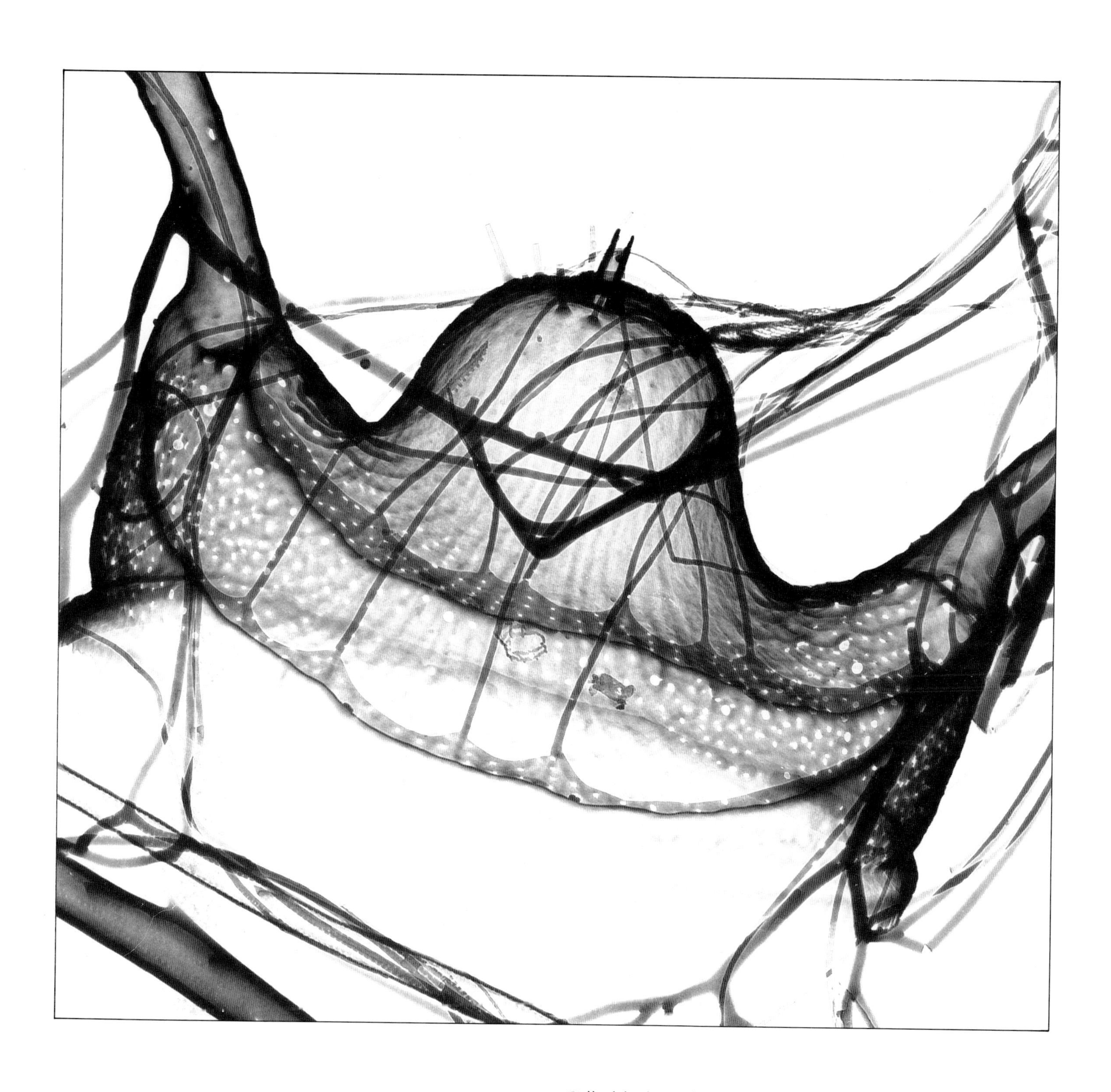

Cell of the hairy diatom *Chaetoceros didymus* (Chaeto = hair; ceros = horn) seen in the transmission electron microscope; Derwent River, Tasmania; width 30 μm.

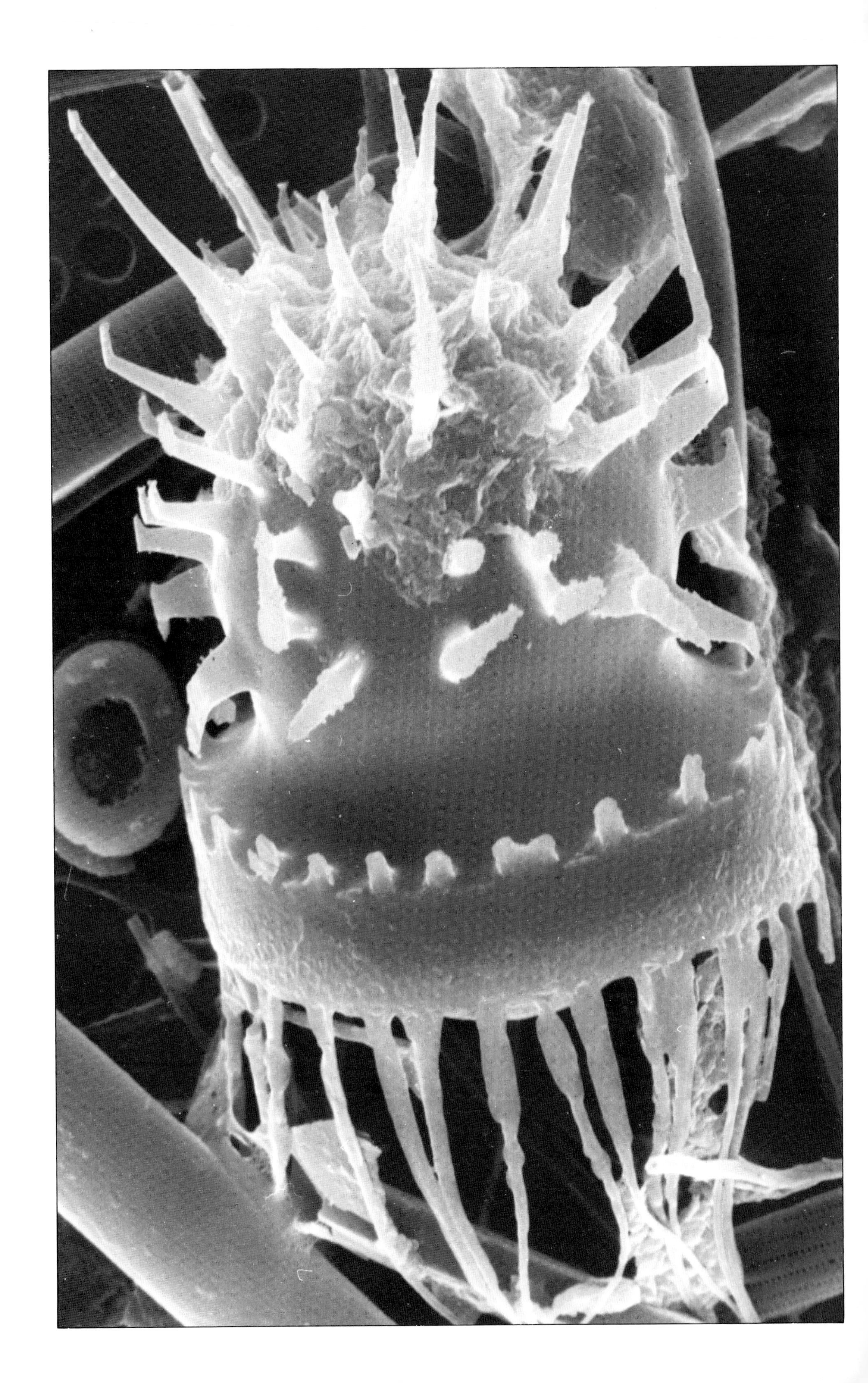

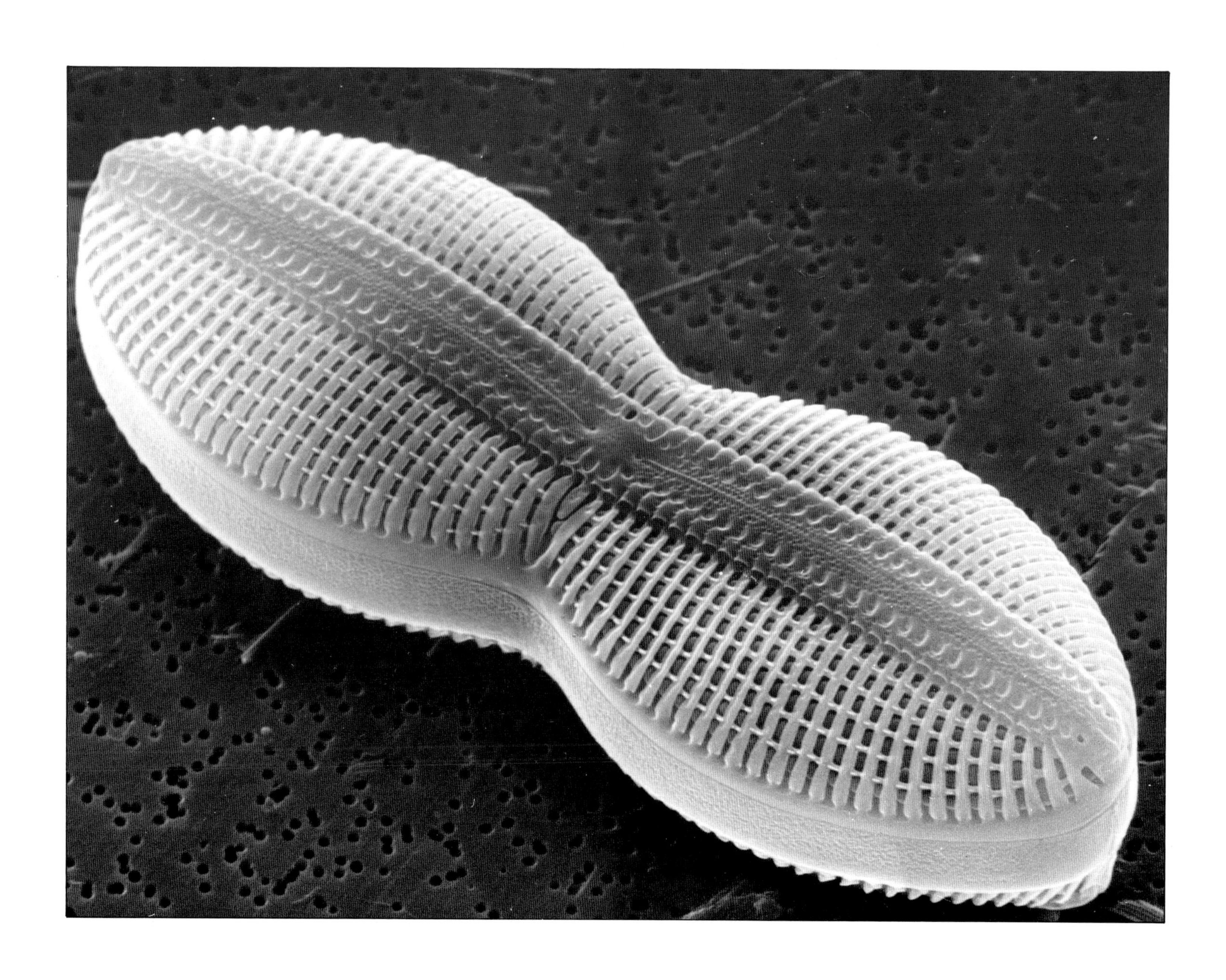

Spined resting spore of the diatom *Chaetoceros lauderi*. These spores form as a response to adverse conditions and under favourable conditions they can germinate to form new diatom cells in the water column; North West Shelf of Australia; length 20 μm.

Footprint-shaped pennate diatom *Diploneis*, which glides on the surface of seaweeds or mudflats; Rottnest Island, Western Australia; length 50 μm.

Strainer-shaped diatom *Cocconeis* which grows attached to seaweeds. Some species of this diatom genus are responsible for the brown coating on whale skins; Rottnest Island, Western Australia; length 30 μm.

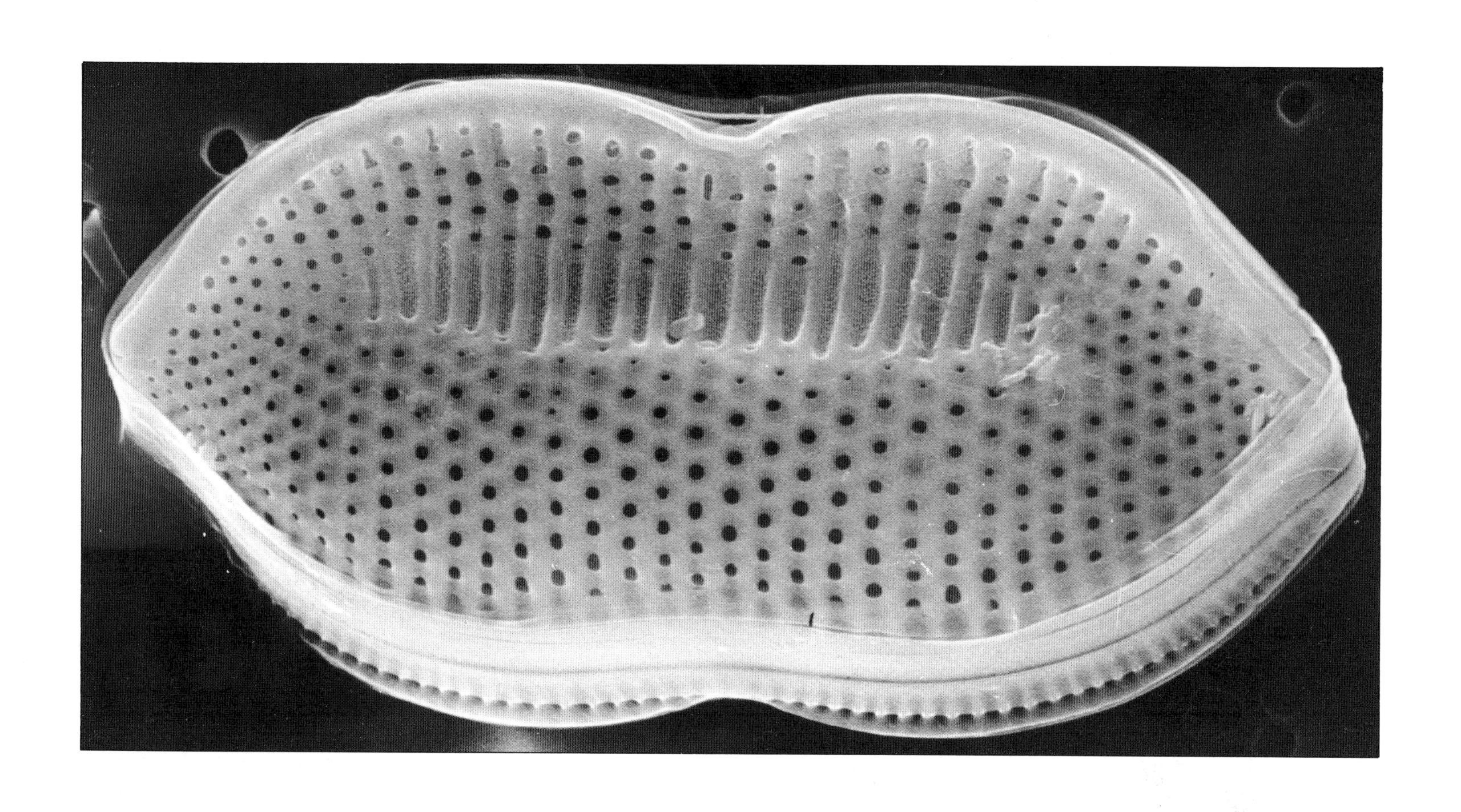

Diatom *Nitzschia panduriformis* (Nitzschia = dedicated to C.L. Nitzsch; panduriformis = violin-shaped); North West Shelf of Australia; length 30 μm.

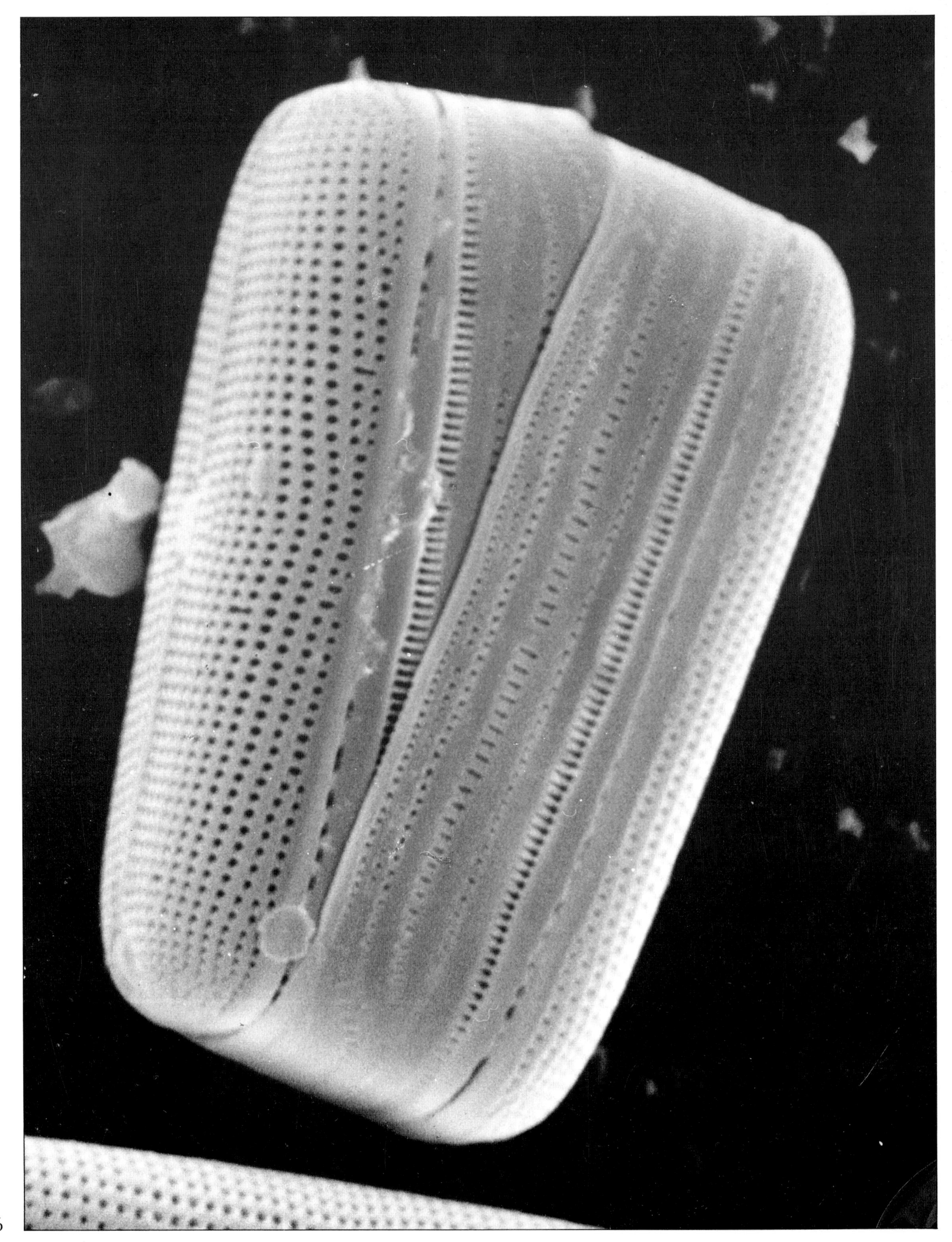

Pill-box shaped diatom *Cistula lorenziana* (Cistula = small box) composed of a lid (epitheca) fitting on a box (hypotheca); Derwent River, Tasmania; length 30 μm.

Zig-zag shaped colony of the diatom *Thalassionema nitzschioides* (Thalassio = sea; nema = thread; nitzschioides = similar to *Nitzschia*). The individual cells attach to each other by mucus pads produced by the cell poles; North West Shelf of Australia; length 30 μm.

Butterfly shaped bottom-dwelling diatom *Campylodiscus*; Huon River, Tasmania; width 50 μm.

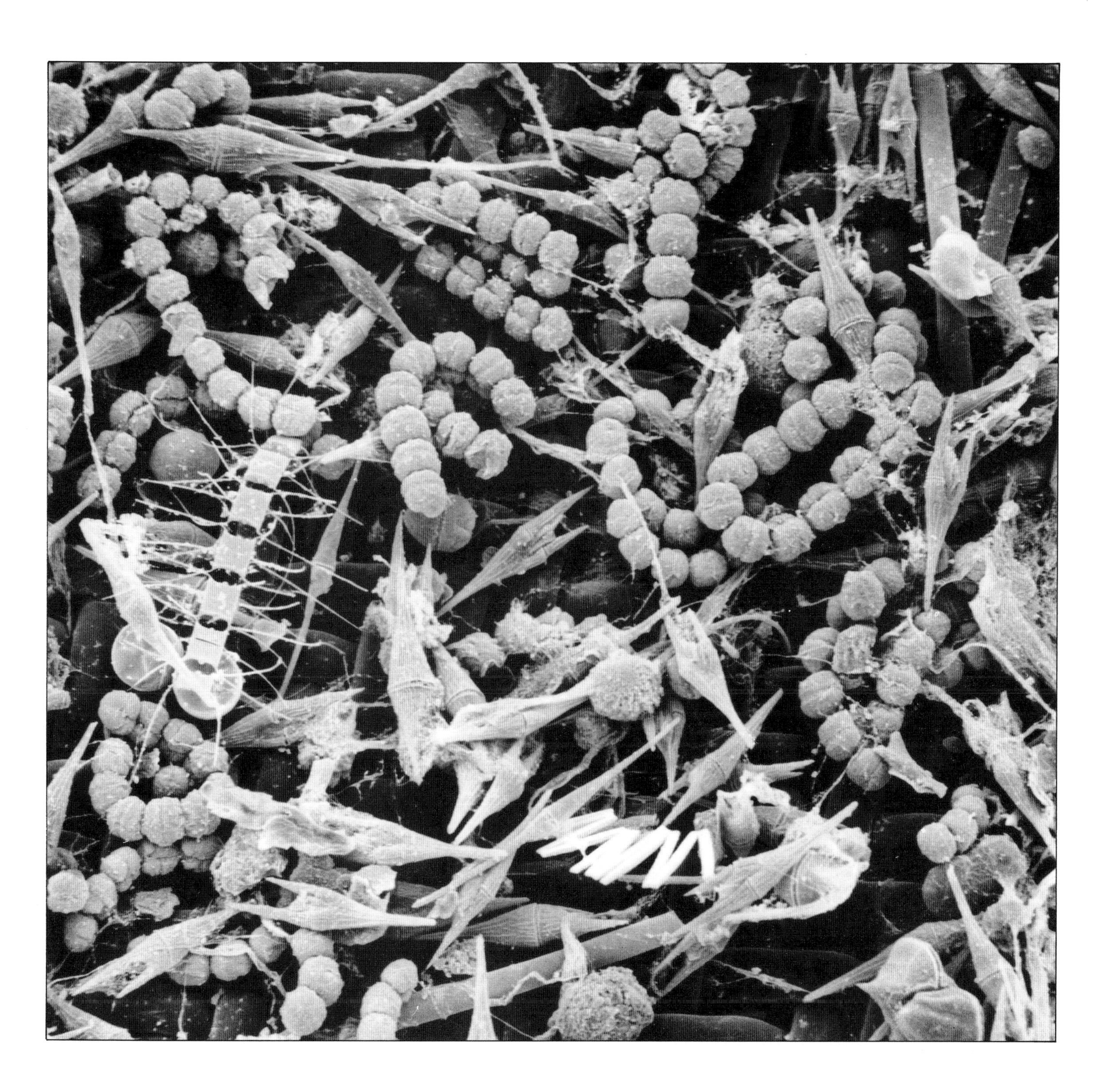

Mixed plankton sample from the River Derwent, Tasmania, dominated by the chainforming toxic dinoflagellate *Gymnodinium catenatum* (bead-like cells). This plankton organism produces potent neurological toxins that can find their way via shellfish to humans where, in extreme cases, they can cause death through respiratory paralysis (paralytic shellfish poisoning, PSP).

Chain-forming, unarmoured (naked) dinoflagellate *Gymnodinium catenatum* (Gymnos = naked; catenatum = chain) bounded by a membranous covering with hexagonal pattern. Each cell has a transverse flagellum for rotational movements and a longitudinal flagellum for forward propulsion; Derwent River, Tasmania; diameter 35 μm.

Unarmoured dinoflagellate *Polykrikos schwartzii* (Poly = many; krikos = rings) in which the individual cells of a chain have fused to form a single large cell. This species is a predator of dinoflagellates such as *Gymnodinium catenatum*; Derwent River, Tasmania; length 50 μm.

Dinoflagellate *Prorocentrum micans* (Pro = on front; centrum = spine). The flattened cell is covered by two large lateral plates ornamented with small depressions and pores; Peel Harvey estuary, Perth; length 45 μm.

Dinoflagellate *Heteroschisma* (Hetero =other; schisma = separation) composed of an upper cell half (epitheca) and lower cell half (hypotheca) separated by a girdle groove; North West Shelf of Australia; diameter 40 μm.

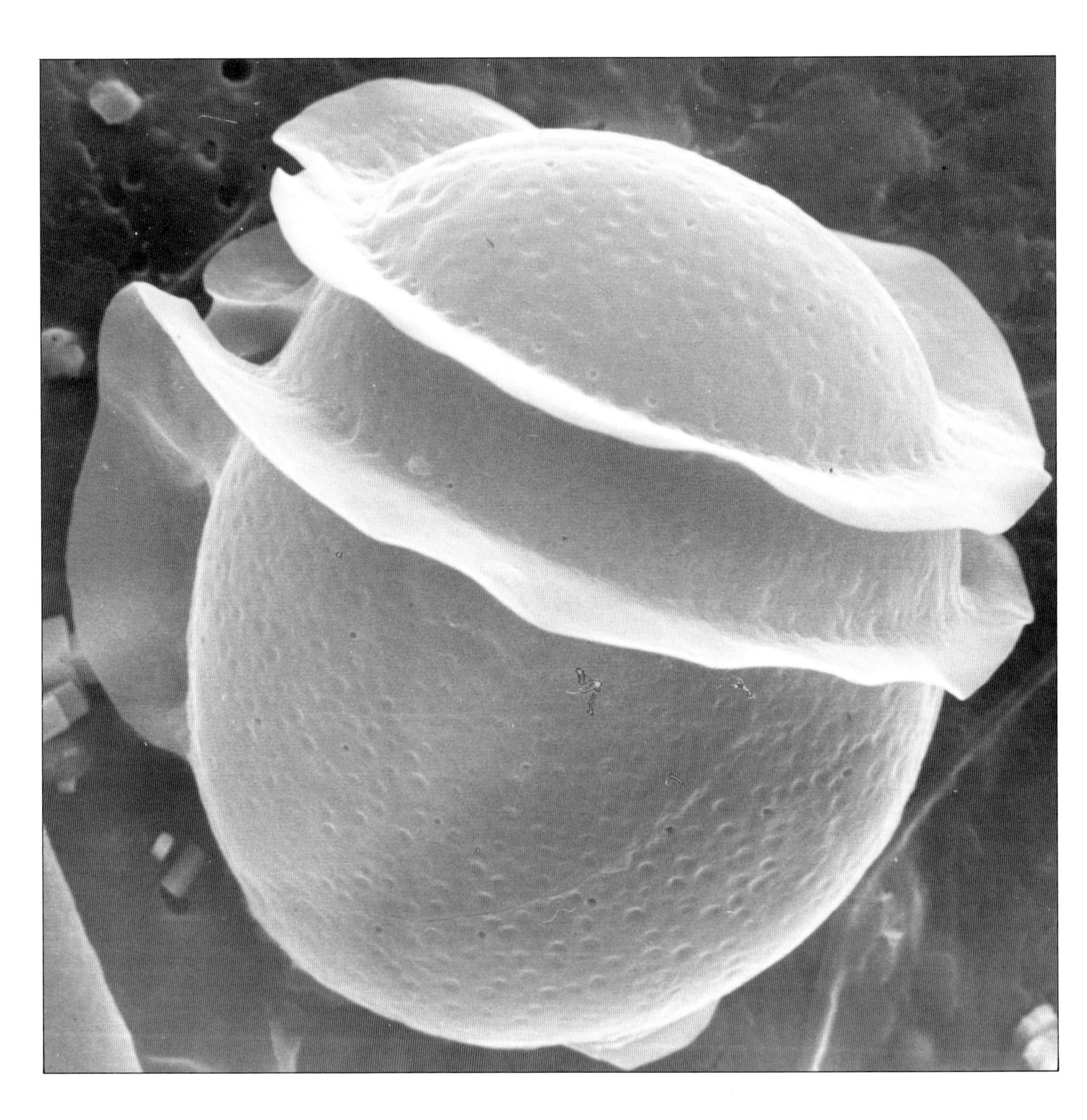

Spherical dinoflagellate *Phalacroma parvulum*, which behaves like an animal and can ingest small algal cells or dissolved organic nutrients; Tasman Sea; diameter 40 μm.

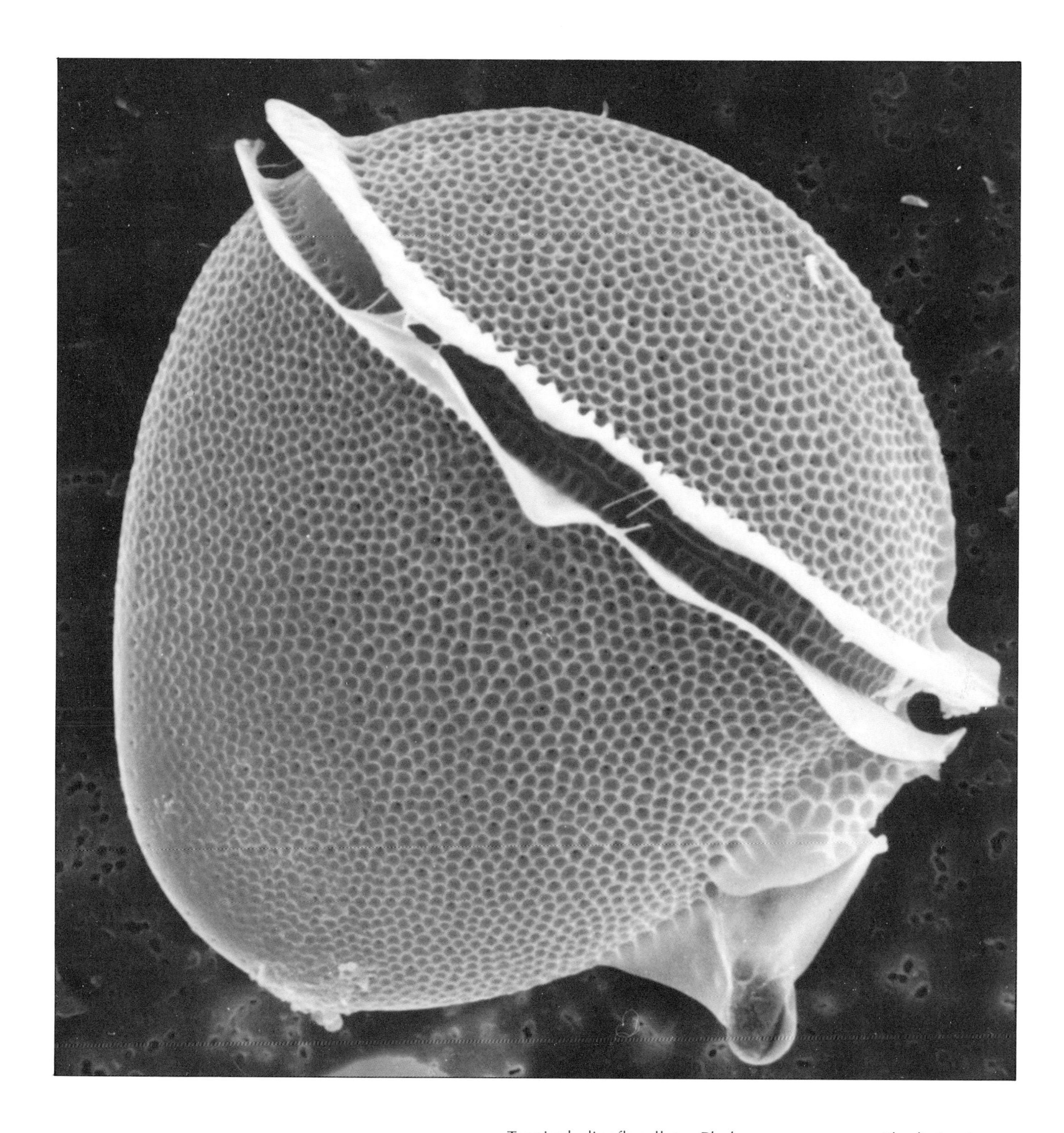

Tropical dinoflagellate *Phalacroma cuneus* with distinctive reticulate ornamentation. This species behaves like a plant and is capable of photosynthesis; North West Shelf of Australia; length 75 μm.

Tropical dinoflagellate *Phalacroma rapa* with heavy areolate ornamentation of the cellulose theca; North West Shelf of Australia; diameter 50 μm.

Strawberry-like dinoflagellate *Dinophysis acuminata* which can cause severe bouts of diarrhoea in humans who eat shellfish contaminated with this plankton species (diarrheic shellfish poisoning, DSP); Derwent River, Tasmania; length 40 μm.

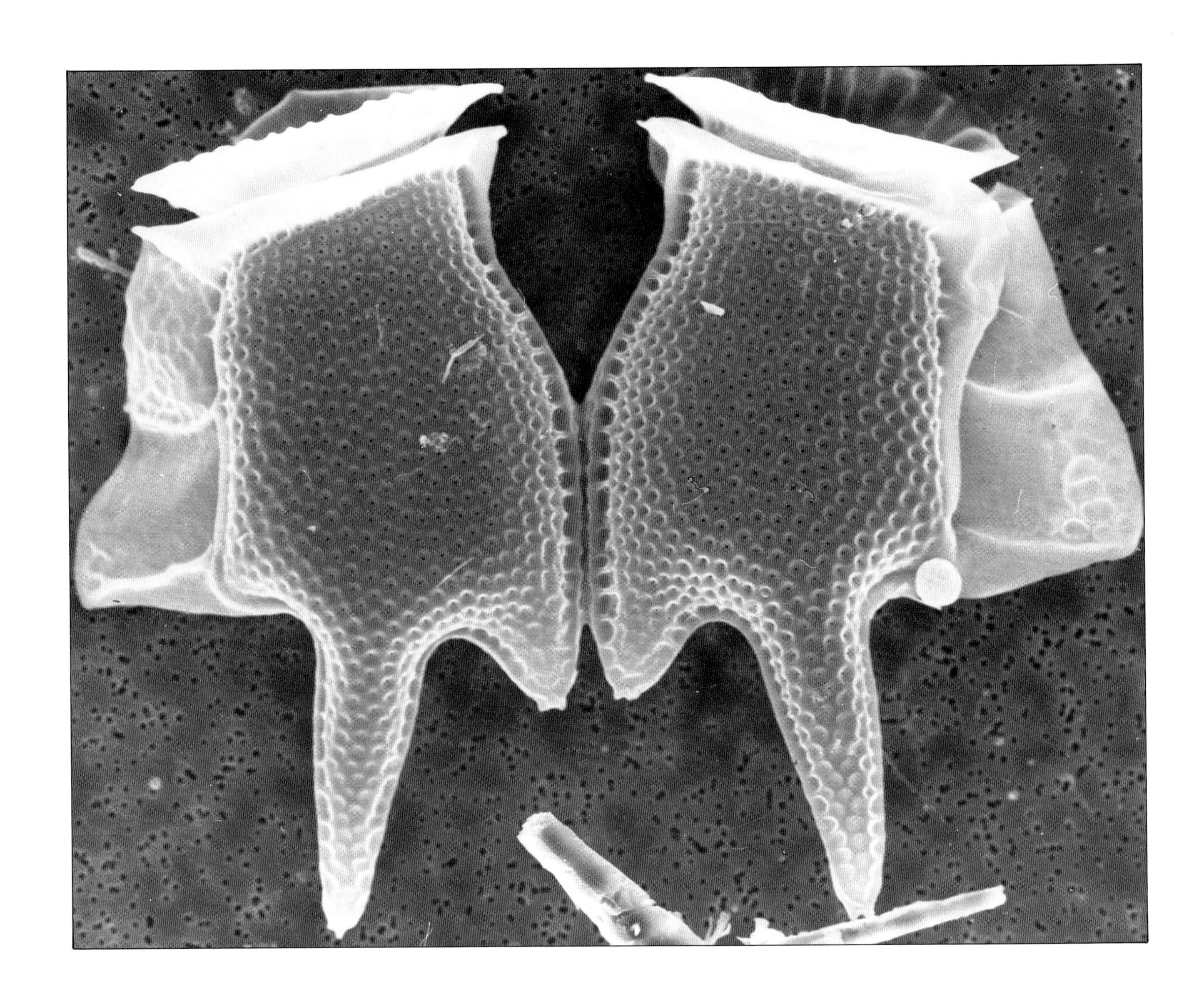

Divided pair of the dinoflagellate *Dinophysis tripos*, still joined by the edges of their hypothecae; Tasman Sea; length 60 μm.

Tropical oceanic dinoflagellate *Dinophysis hastata* with distinctive pointed spine on the hypotheca and a wide, wing-like appendage (list); Indian Ocean; length 50 μm.

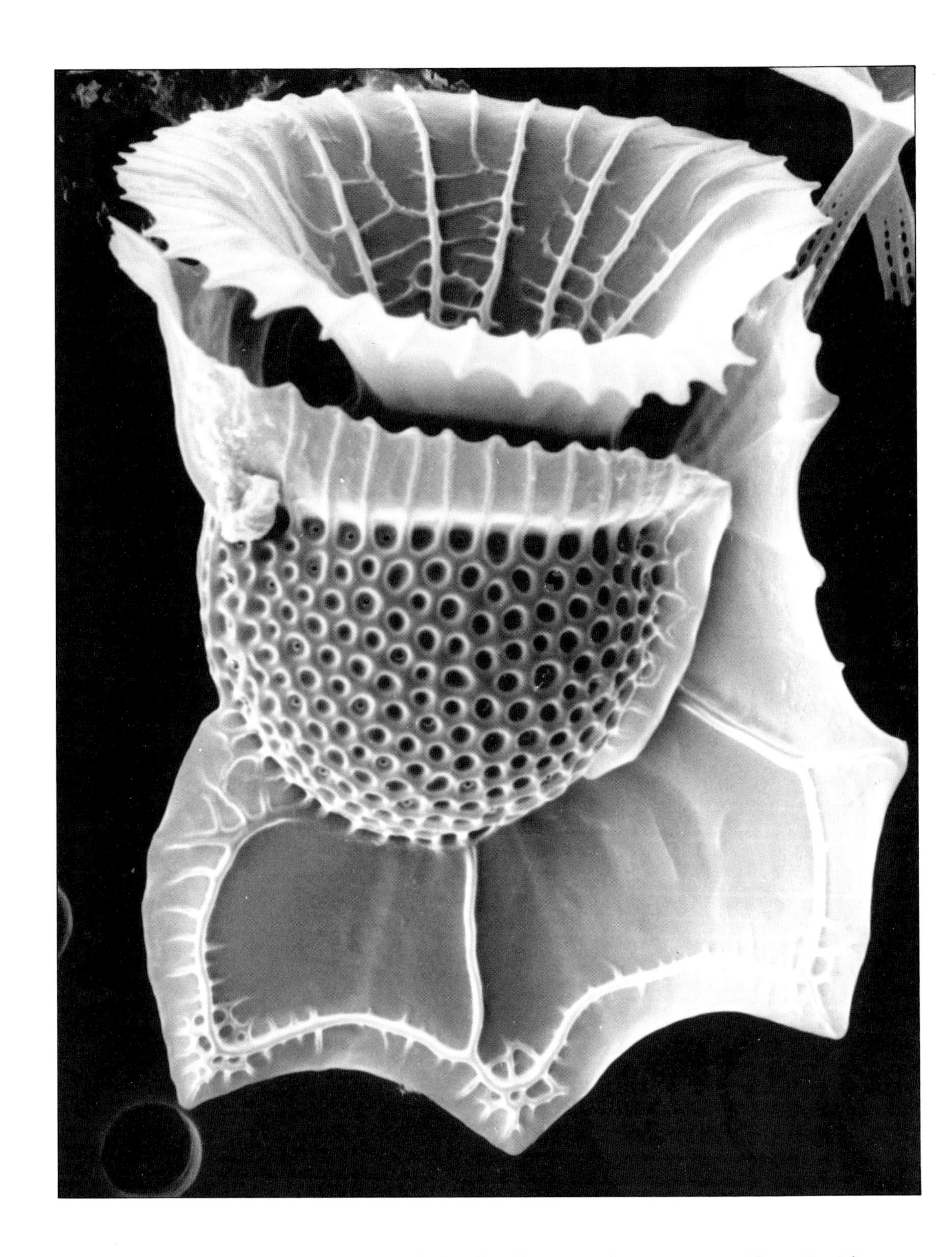

Tropical dinoflagellate *Ornithocercus magnificus* (Ornitho = bird; cercus = tail) with parachute-like wings that are thought to function as a flotation device; North West Shelf of Australia; length 80 μm.

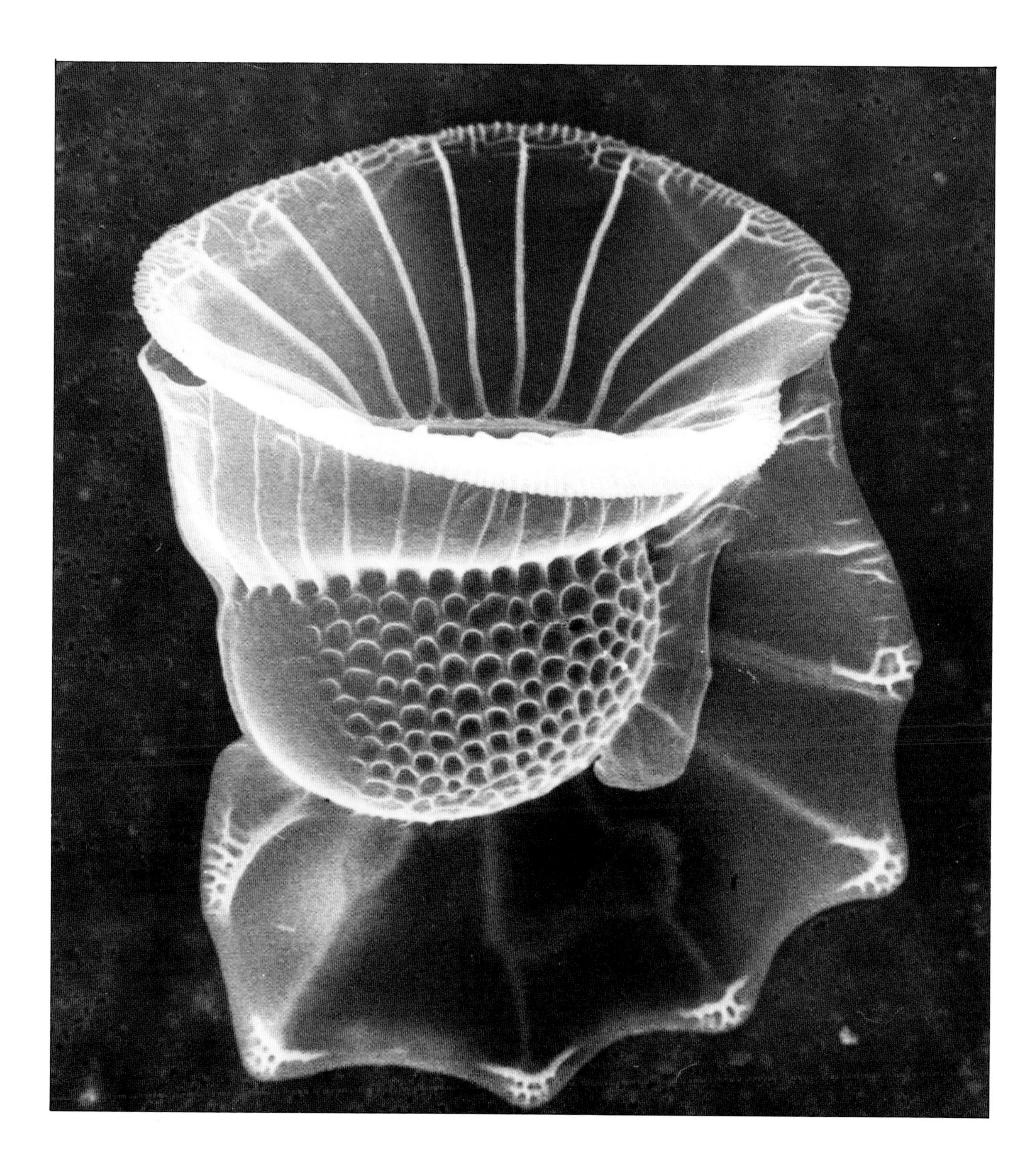

Tropical dinoflagellate *Ornithocercus steinii* which has more extensively lobed lists than *Ornithocercus magnificus*; Gulf of Carpentaria; length 70 μm.

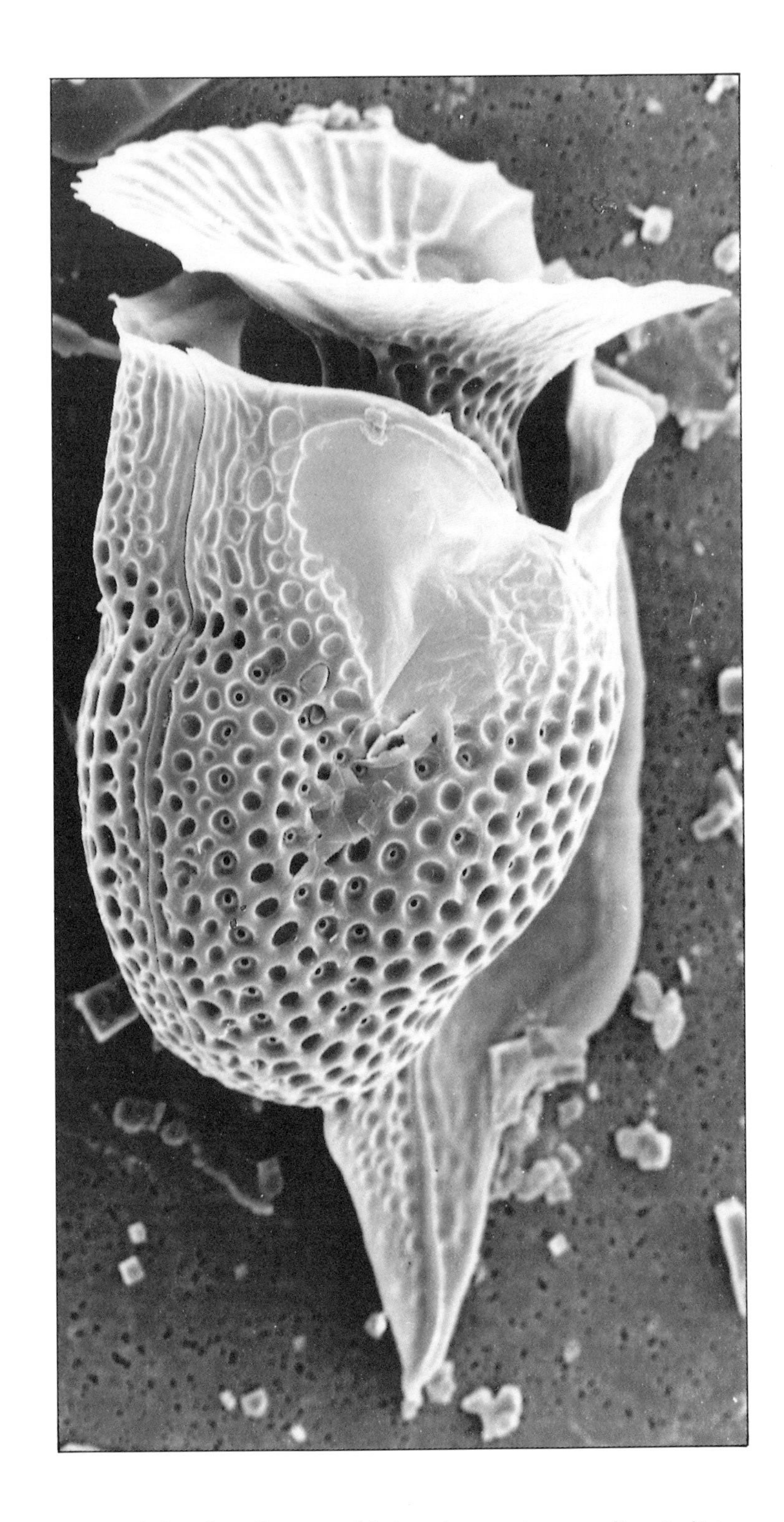

Tropical dinoflagellate *Parahistioneis para,* intermediate in list development between *Ornithocercus* and *Histioneis*; Coral Sea; length 60 μm.

Tropical dinoflagellate *Histioneis mitchellana* (Histion = wing; neis = vessel) with reticulated wings; Indian Ocean; length 60 μm.

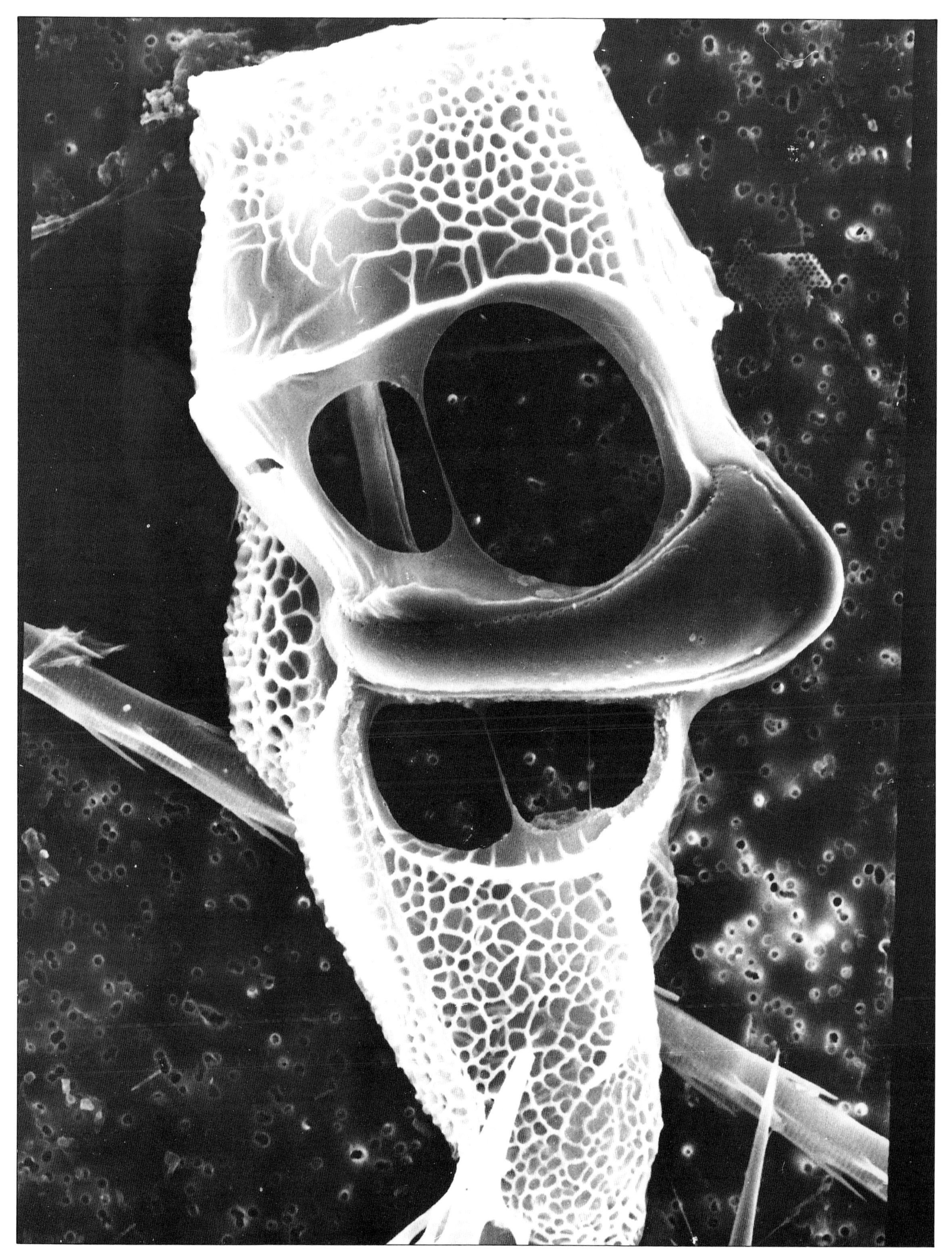

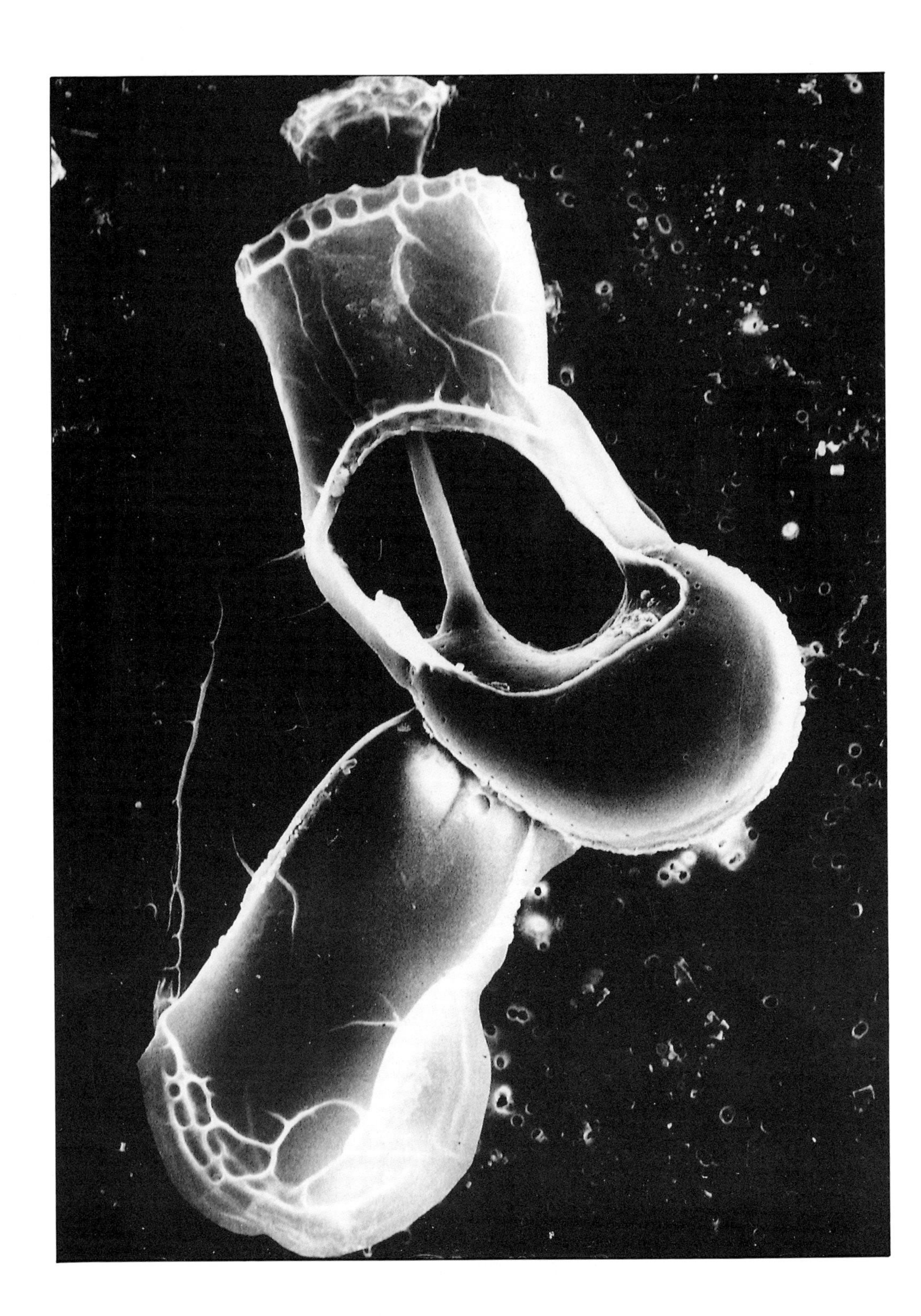

Unusually shaped tropical dinoflagellate *Histioneis dolon*. The central chamber functions as a greenhouse in which this animal-like organism grows blue-green algal symbionts to supplement its diet; Indian Ocean; length 70 μm.

Needle-shaped dinoflagellatė *Amphisolenia bidentata* with reduced epitheca (head) and extremely elongated hypotheca (foot); East Australian Current; 750 μm long.

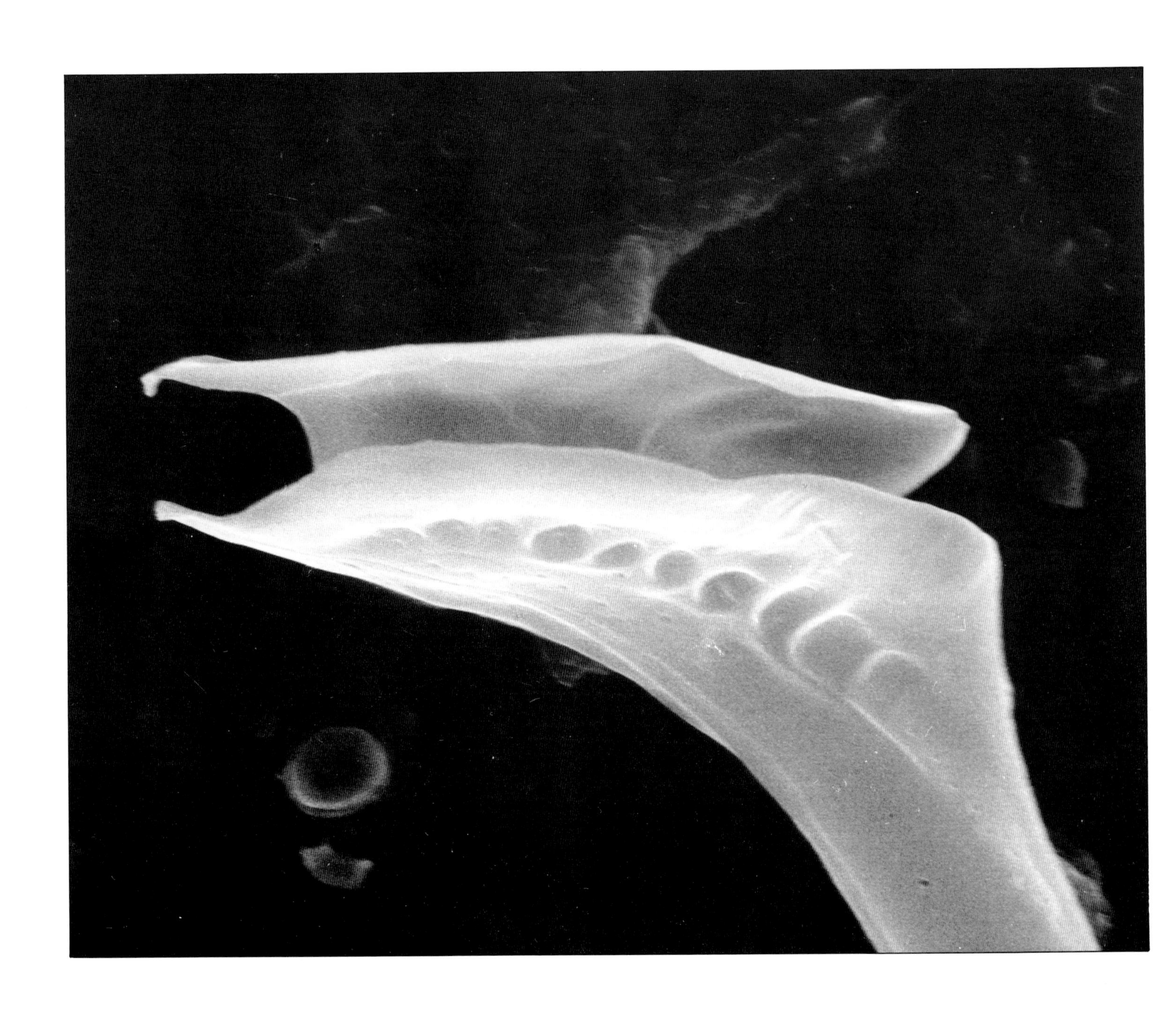

Detail of the epitheca (head) of *Amphisolenia bidentata*.

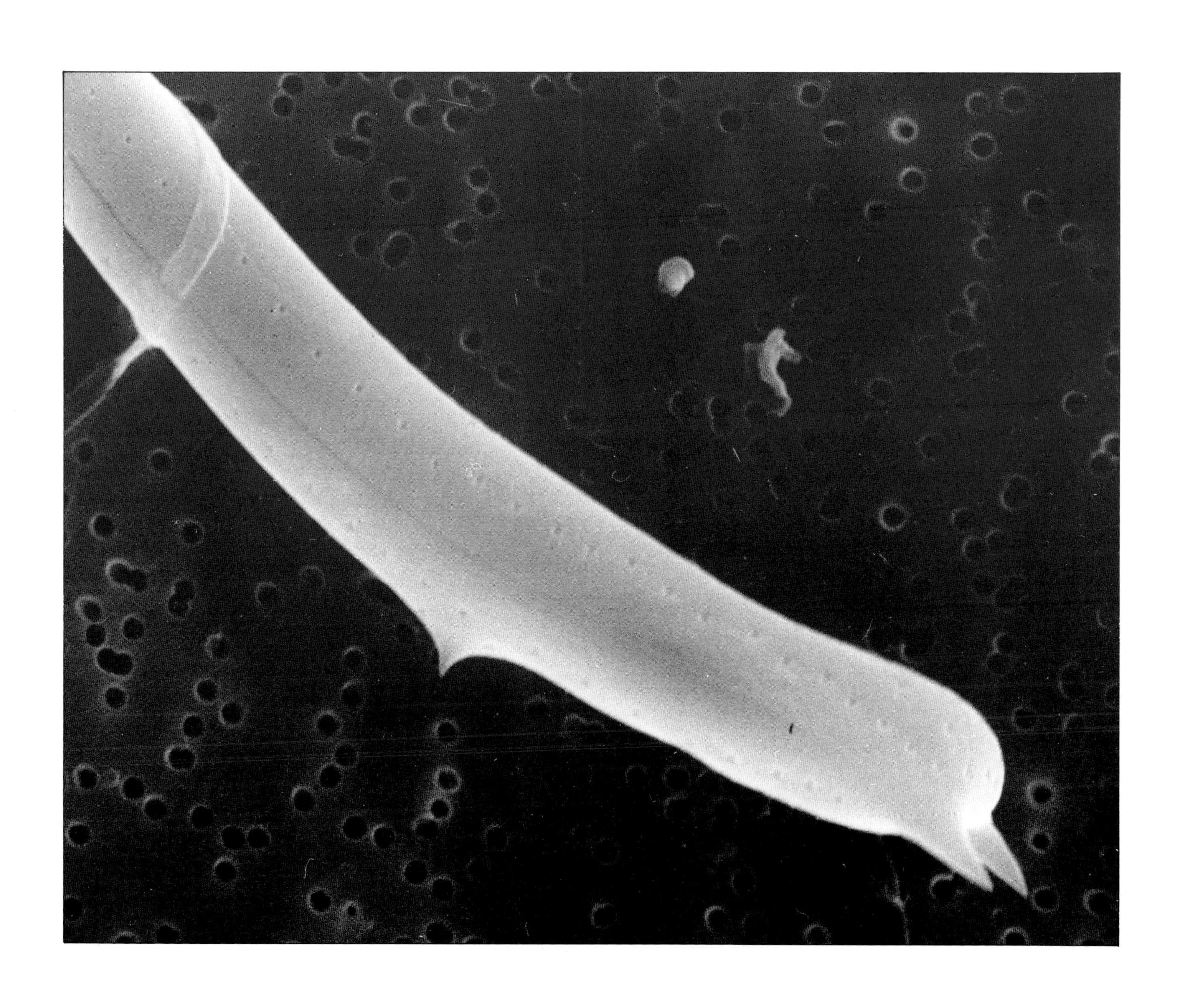

Detail of the foot end of *Amphisolenia bidentata,* showing the two distinctive spines for which this species is named.

Lateral view of the dinoflagellate *Goniodoma polyedricus* with distinctive angular ridges on the thick cellulose cell wall; East Australian Current; diameter 60 μm.

Top view of the dinoflagellate *Goniodoma polyedricus* showing a radially symmetric plate pattern with three polar plates.

Bottom view of the dinoflagellate *Goniodoma polyedricus* showing a radially symmetric plate pattern with three polar plates.

Dinoflagellate *Gonyaulax spinifera* (Gonyo = angular; aulax = groove; spinifera = spine bearing) with heavily reticulated cellulose wall.

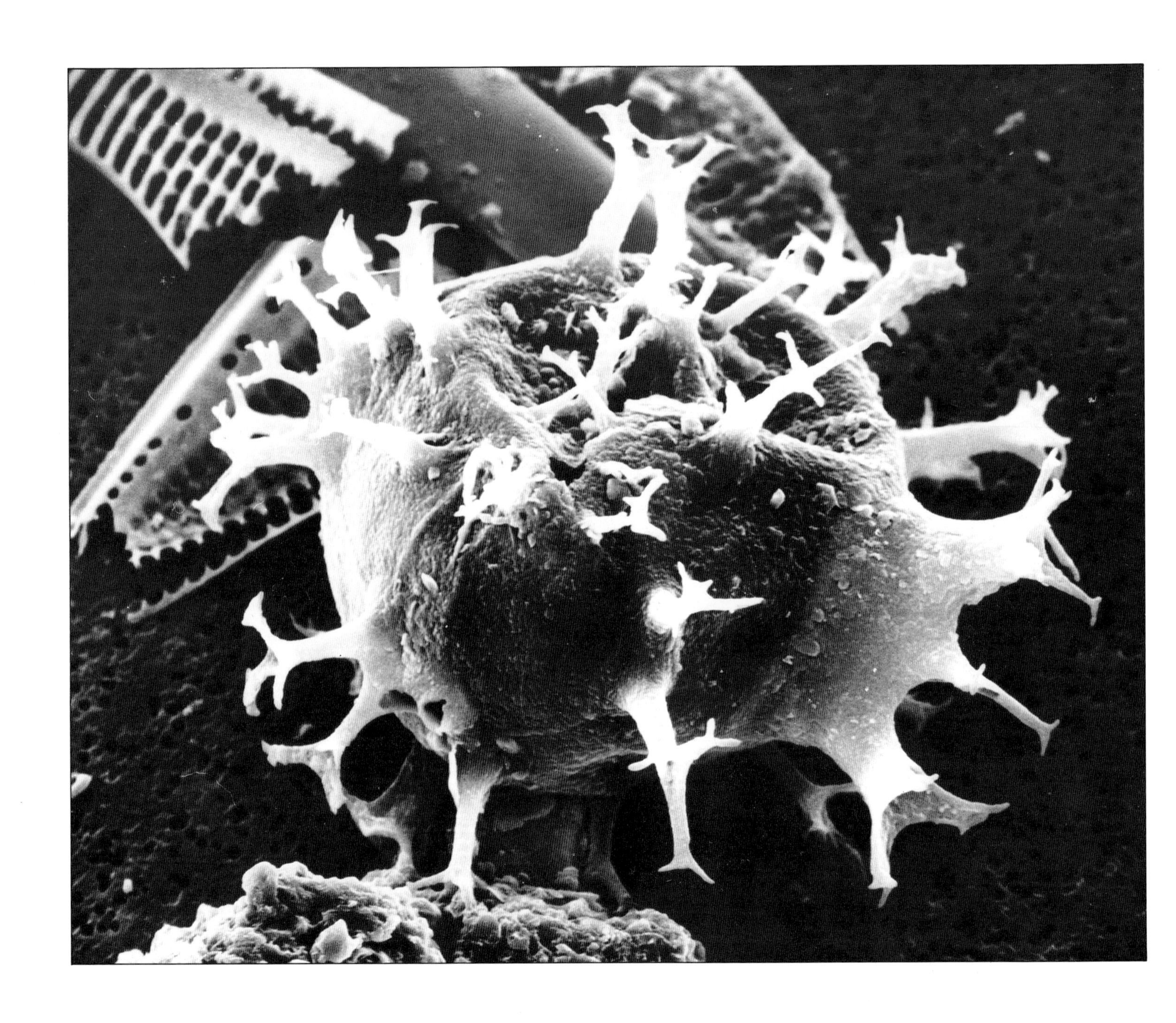

As part of its life cycle the dinoflagellate *Gonyaulax spinifera* species produces a resistant spinose resting cyst with sporopollenin wall known to geologists as *Spiniferites mirabilis*; Derwent River, Tasmania; dinoflagellate cell diameter 40μm.

Beautifully ornamented globular dinoflagellate *Gonyaulax grindleyi*; Huon River, Tasmania; diameter 47 μm.

Ovoid spiny resting cyst of the dinoflagellate *Gonyaulax grindleyi,* known to geologists as *Operculodinium centrocarpum.*

The toxic, tropical, chain-forming dinoflagellate *Pyrodinium bahamense* (Pyro = fire; bahamense = first described from the Bahamas). This species has been responsible for about 100 fatal cases of paralytic shellfish poisoning in the Indo-West Pacific and Caribbean; Port Moresby, Papua New Guinea; diameter 50 μm.

A single cell of the dinoflagellate *Pyrodinium bahamense* seen in lateral view.

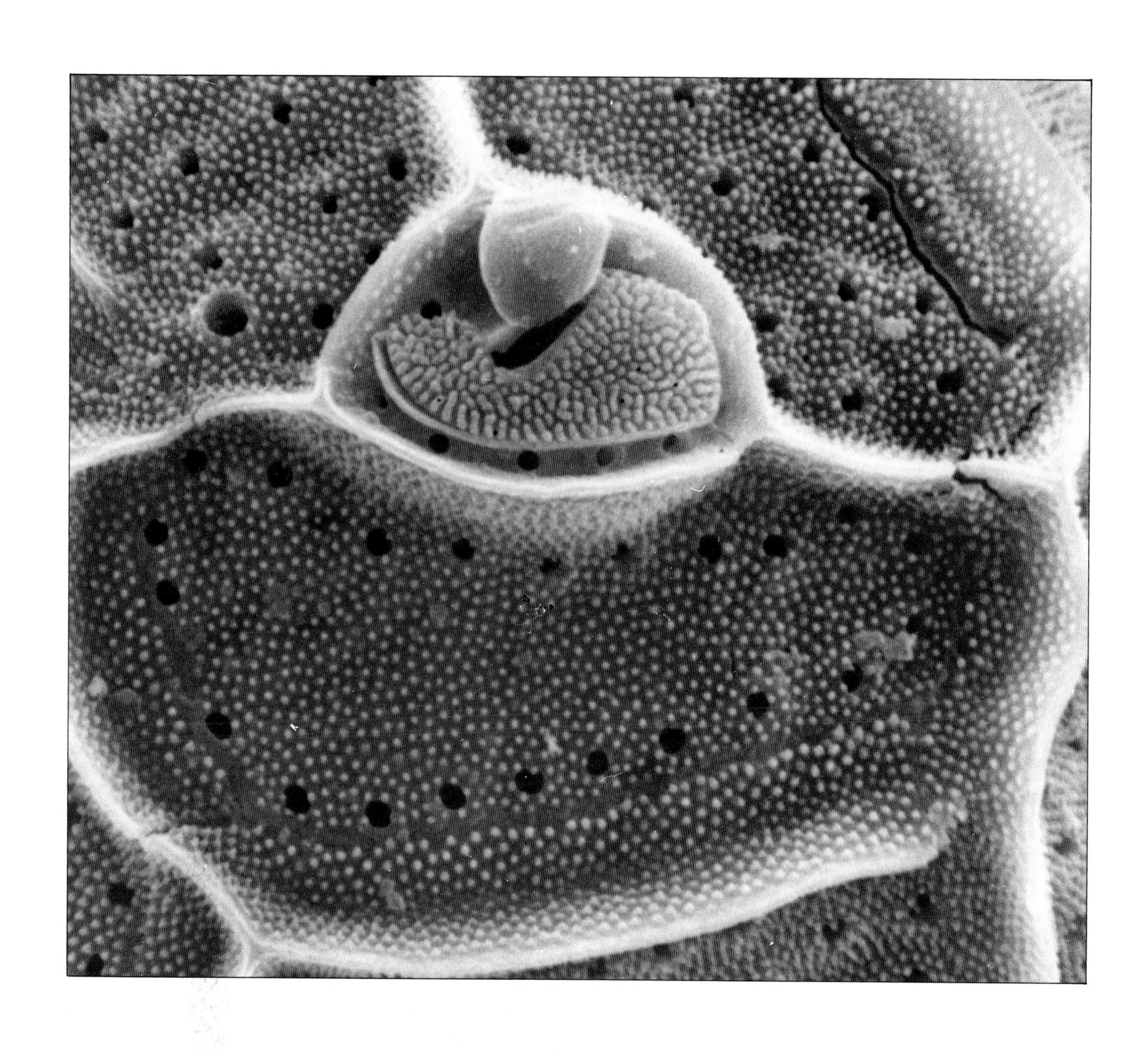

Detail of thecal ornamentation and apical pore of the dinoflagellate *Pyrodinium bahamense*.

Heavily reticulated tropical dinoflagellate *Protoceratium spinulosum*; Coral Sea; diameter 35 μm.

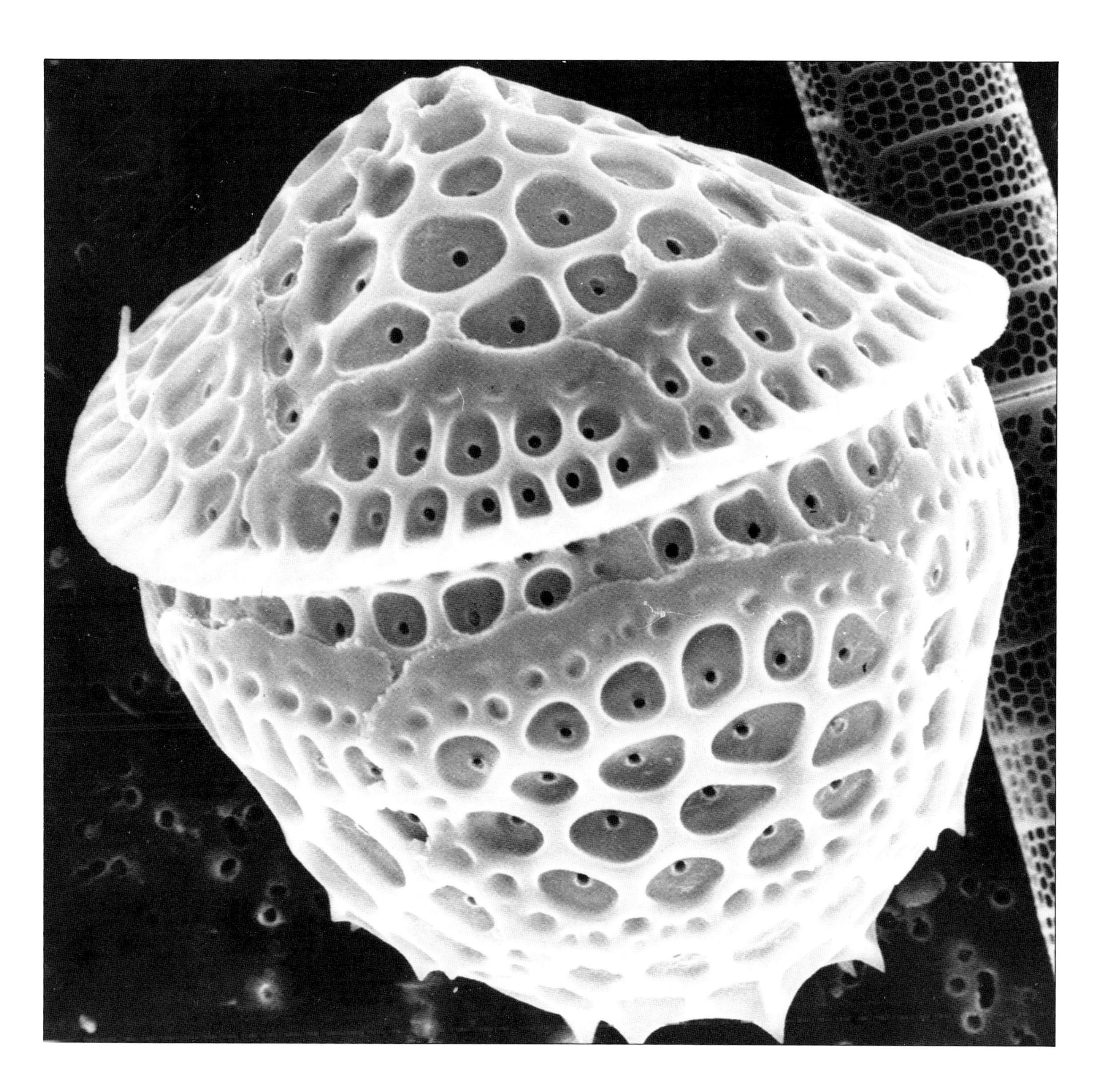

Tropical dinoflagellate *Heterodinium milneri* with distinctive reticulations each enclosing a central pore; Indian Ocean; diameter 60 μm.

Biconical tropical dinoflagellate *Gonyaulax jollifei* with thick thecal plates and strong horns. This species is indicator of tropical oceanic waters; Coral Sea; length 55 μm.

Leaf-like tropical dinoflagellate *Heterodinium whittingae* from 150m depth ('shade flora') in the Coral Sea; length 100 μm.

Tropical dinoflagellate *Ceratocorys horrida* (Cerato = horn; corys = helmet) which looks like a gladiator's helmet. The horn-like extensions are thought to function as a defence mechanism against grazing by small plankton animals; Gulf of Carpentaria; length 100 μm.

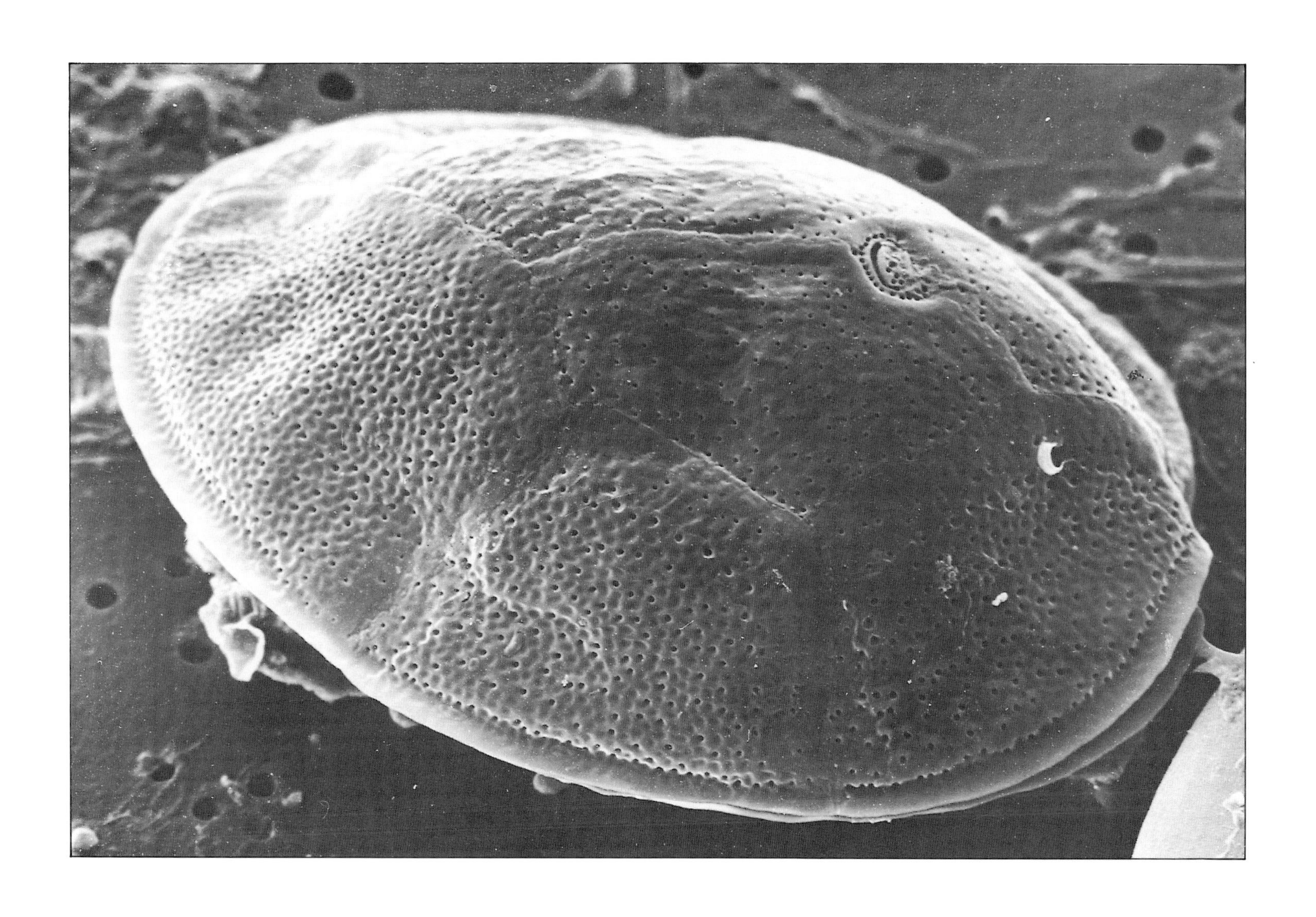

Bottom - dwelling dinoflagellate *Gambierdiscus toxicus* (Gambier = first described from the Gambier Islands) from coral rubble from the Great Barrier Reef. This species produces potent neurotoxins that can find their way from small fish grazing on the reef into big carnivorous fish to humans where, in extreme cases, they can cause death through respiratory failure (ciguatera fish poisoning); 60 μm.

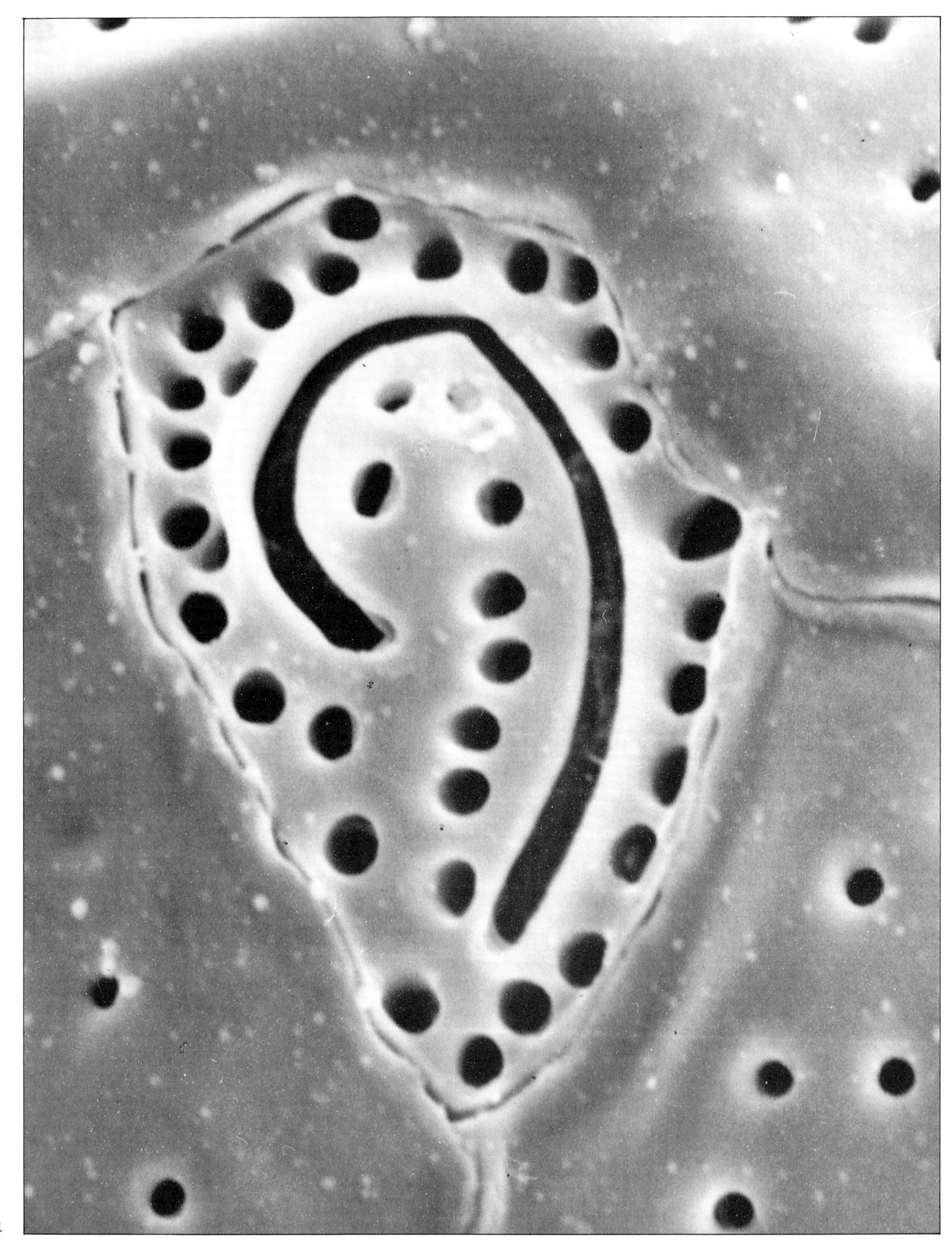

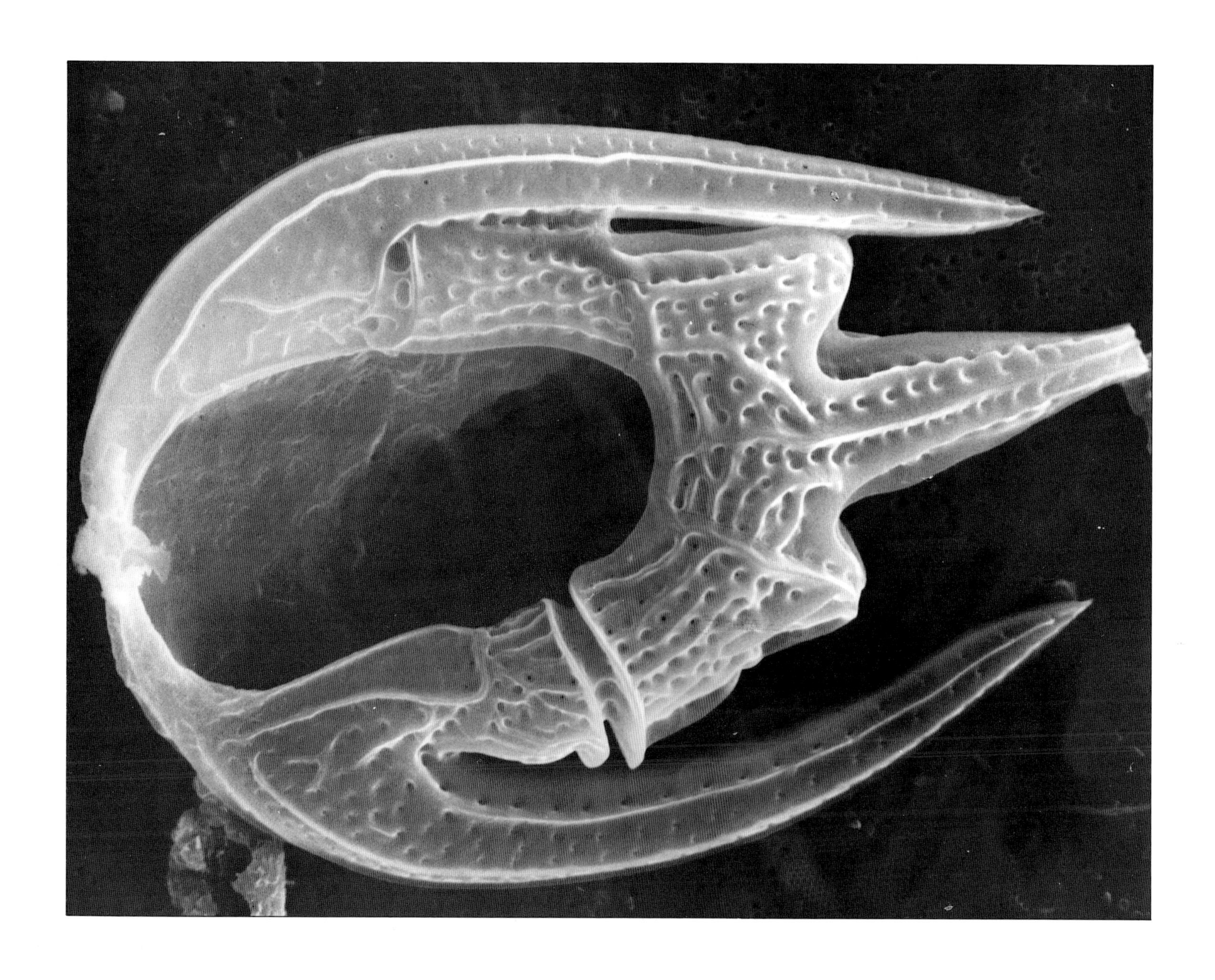

Detail of apical pore on top of the flattened cell of the dinoflagellate *Gambierdiscus toxicus,* showing hieroglyphe-like ornamentation.

Anchor-shaped dinoflagellate *Ceratium paradoxides* (Ceratium = little horn). The opening on front of the cell is thought to allow ingestion of prey; Coral Sea; length 70 μm.

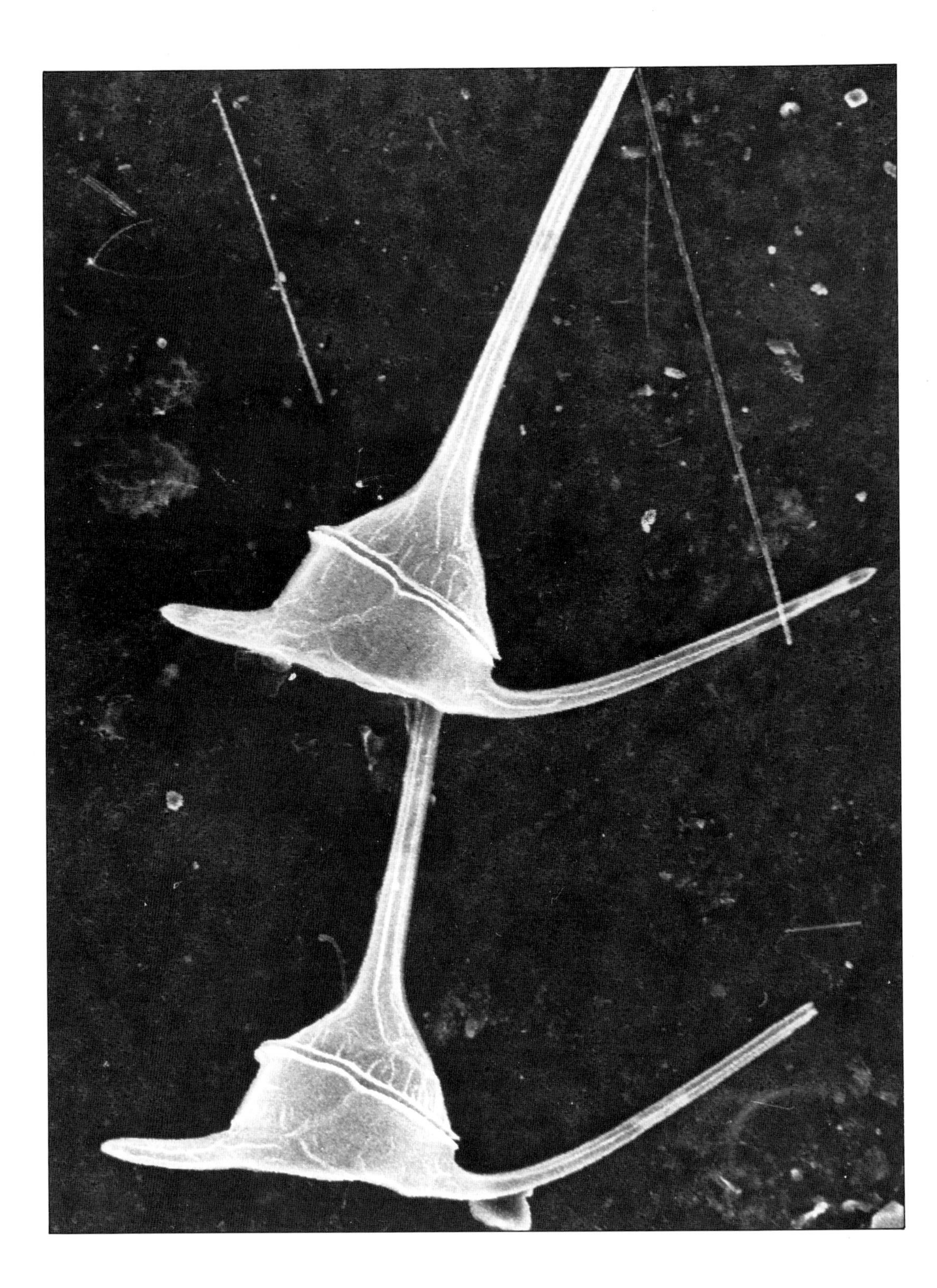

Chain of two cells of the dinoflagellate *Ceratium dens*. This species is found only in tropical coastal waters in the Indian Ocean; length 80 μm.

The biconical dinoflagellate *Oxytoxum constrictum* (Oxy = pointed; toxon = arc); East Australian Current; length 45 μm.

Minute dinoflagellate *Oxytoxum laticeps,* which is an important component of the nanoplankton (dwarf plankton) in the Coral Sea; length 12 μm.

Unusually shaped dinoflagellate *Podolampas bipes* (Podo = foot; lampas = torch; bipes = two feet) which lacks the transverse groove found in most other dinoflagellates; North West Shelf of Australia; length 100 μm.

Small colourless dinoflagellate *Protoperidinium brevipes* (brevipes = short feet) with distinctive thecal ornamentation; River Derwent, Tasmania; length 24 μm.

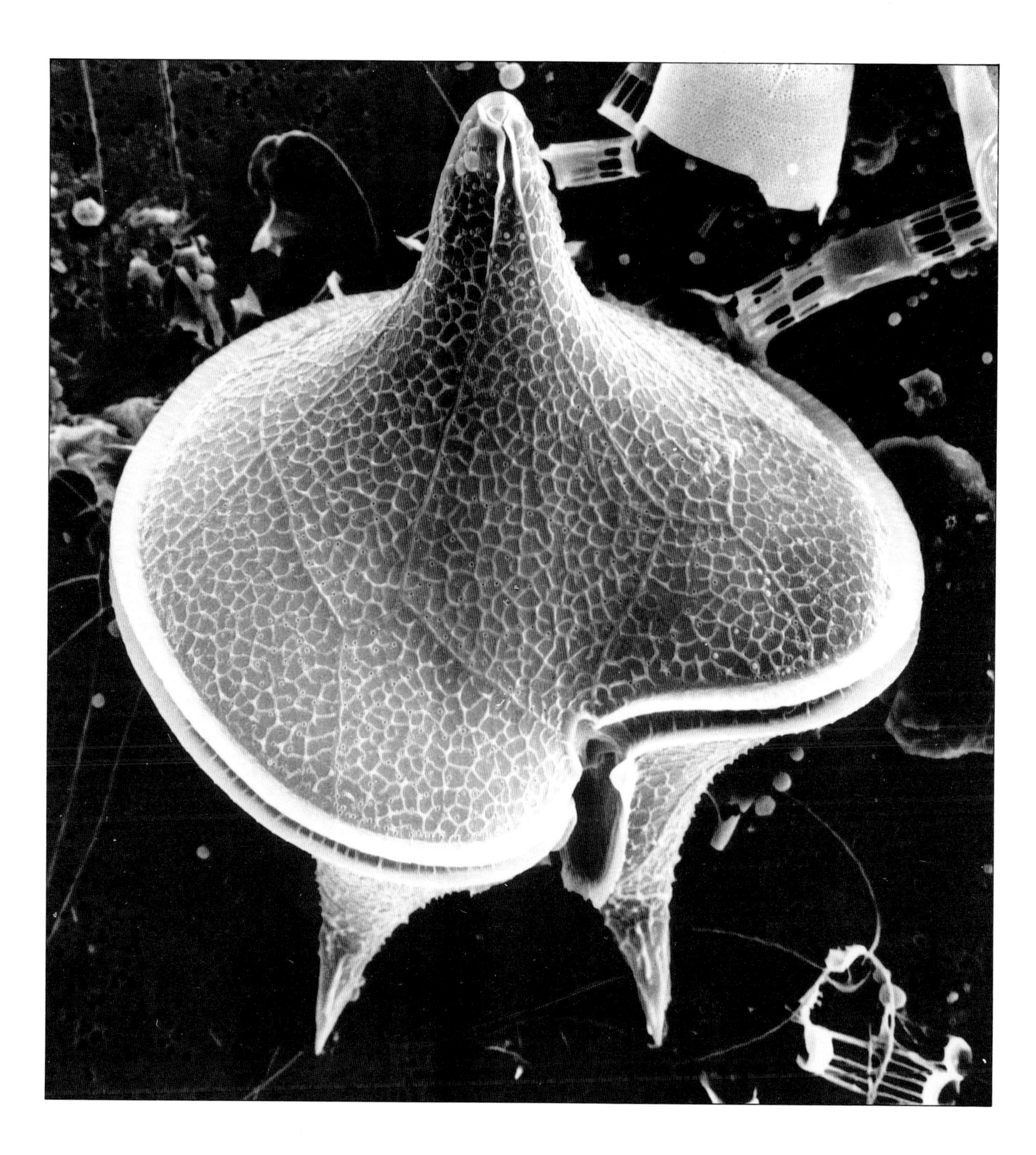

Large dinoflagellate *Protoperidinium divergens* with two diverging antapical horns; Storm Bay, Tasmania; length 70 μm.

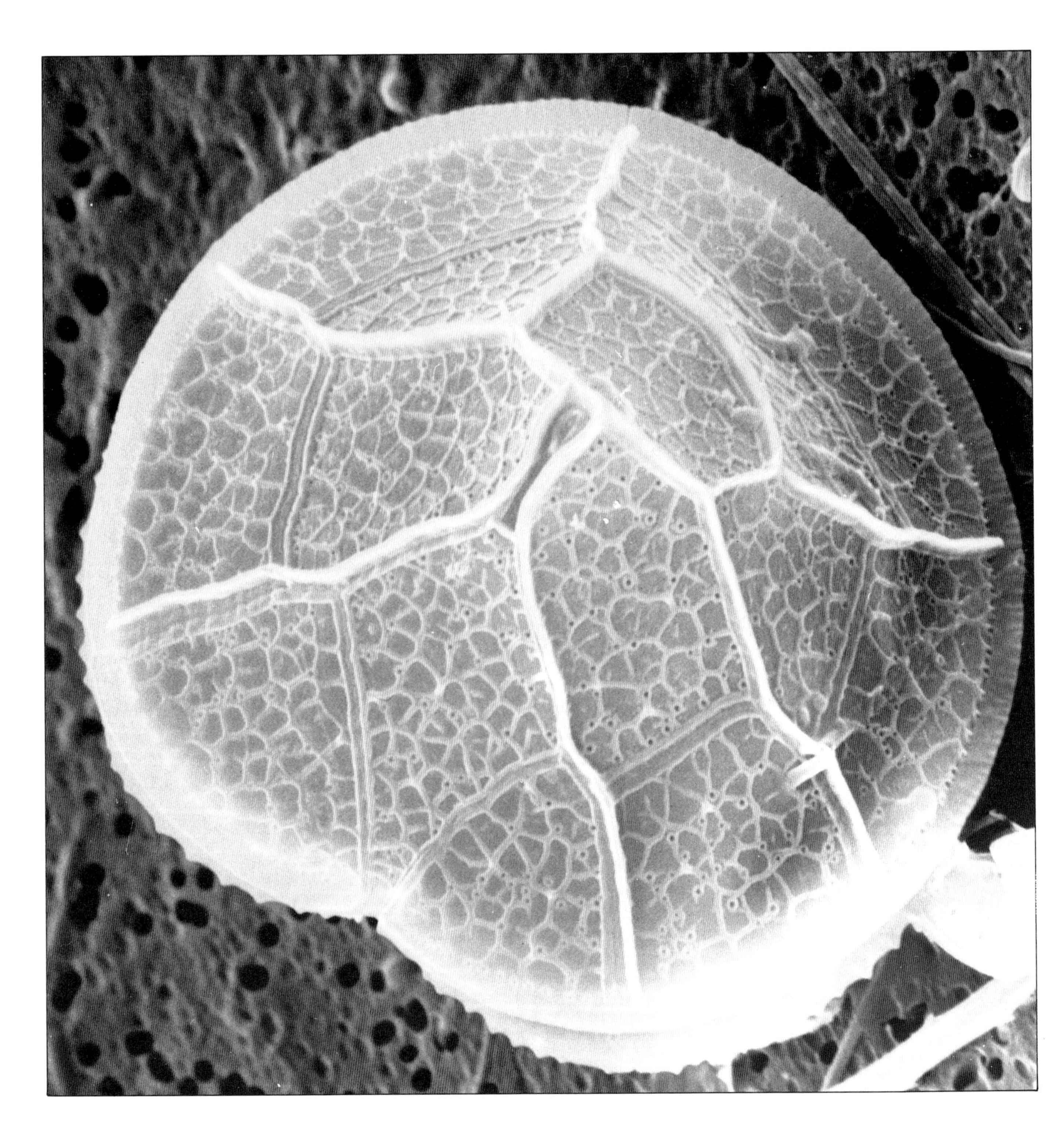

Top view of the dinoflagellate *Protoperidinium leonis* with the epitheca composed of fourteen cellulose plates that fit together like jigsaw pieces; Derwent River, Tasmania; length 70 μm.

Detail of cellulose armour of the dinoflagellate *Protoperidinium leonis*. The horizontal (girdle) groove guides the transverse flagellum and the vertical (sulcus) groove guides the longitudinal flagellum. The cell wall is decorated with a complex pattern of pores, spines and ridges.

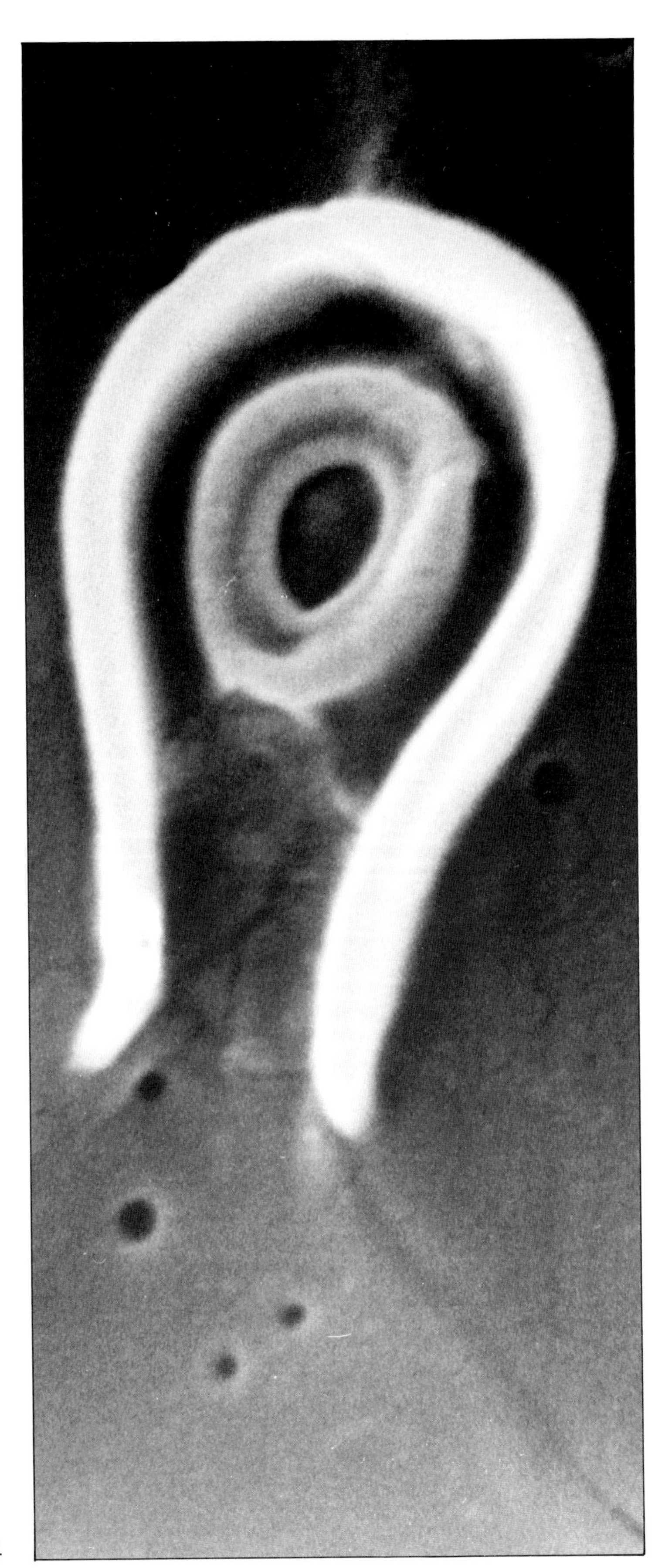

Detail of key-hole shaped pore on top of a large *Protoperidinium* cell.

Small spherical dinoflagellate *Palaeophalacroma unicinctum* which is related to *Protoperidinium* and not *Phalacroma* as the name erroneously suggests; Tasman Sea; length 35 μm.

Egg-shaped dinoflagellate *Blepharocysta splendor-maris* (Blepharo = eyelid; cysta = vessel) which, like *Podolampas,* does not have a transverse girdle groove; North West Shelf of Australia; diameter 60 μm.

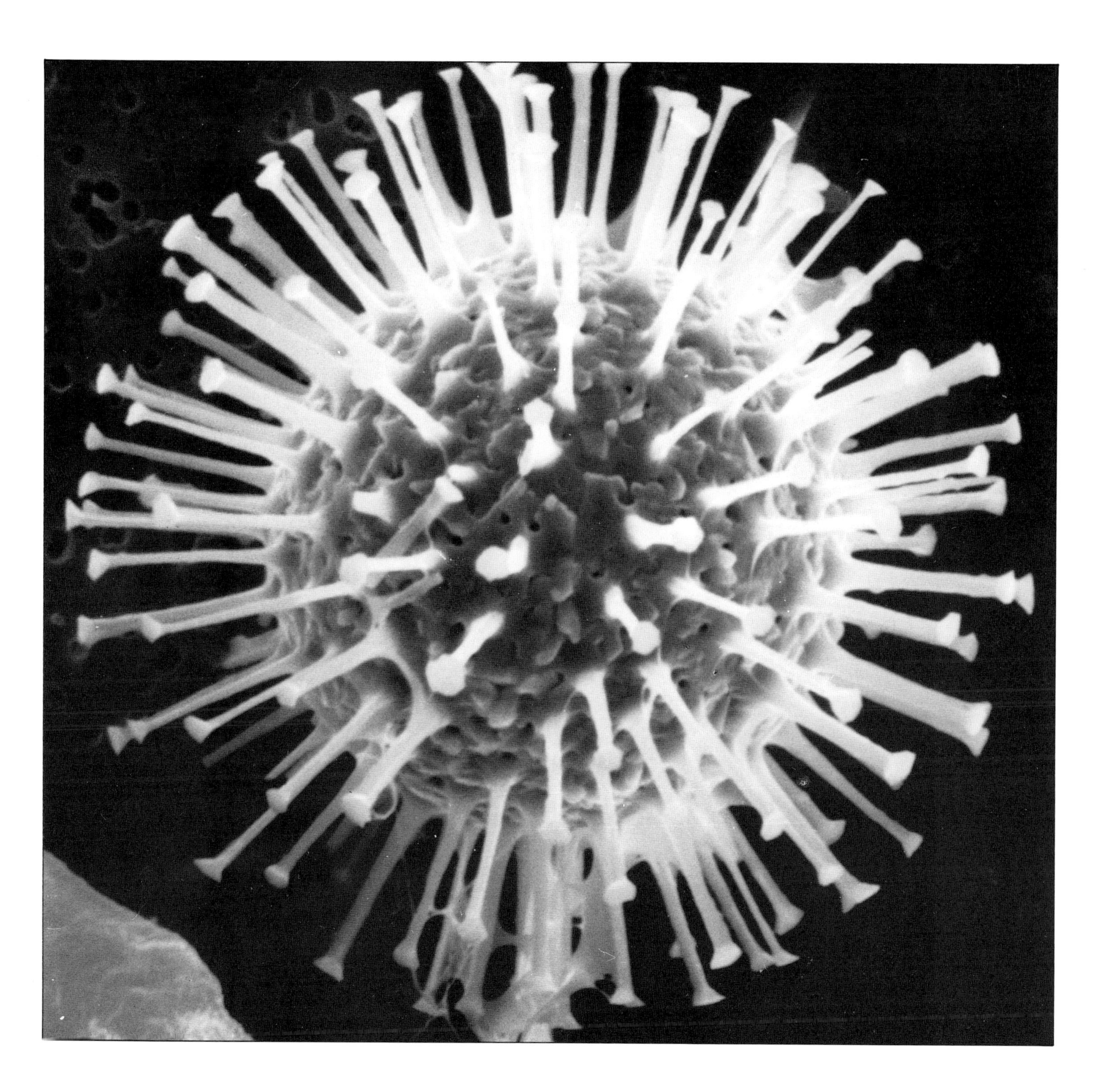

Spinose dinoflagellate cyst known to geologists as *Rhabdothorax regale*; Coral Sea plankton; diameter 40 μm.

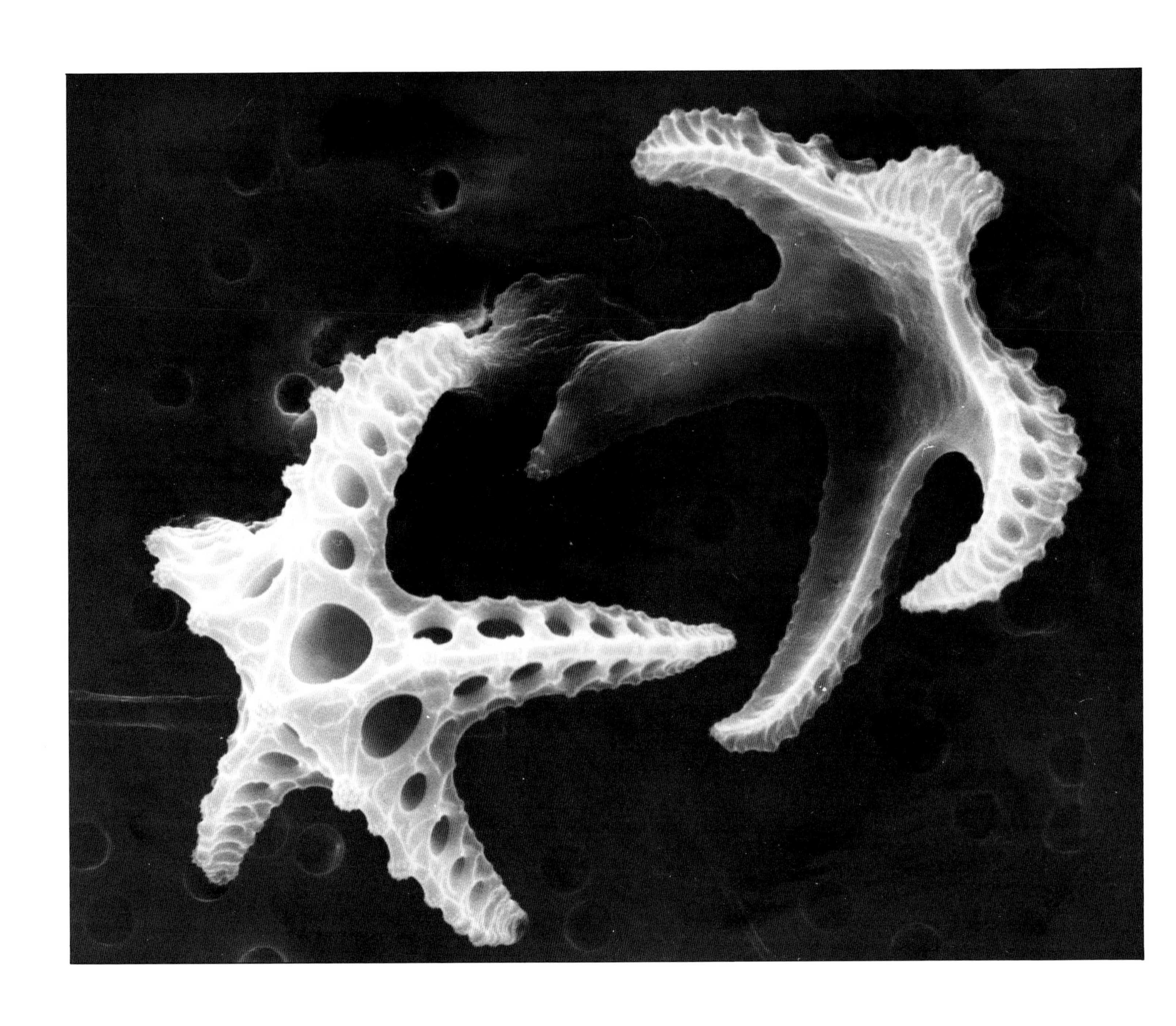

Star-shaped, siliceous internal skeleton of the naked dinoflagellate *Actiniscus*. These structures have been recovered from fossil sediments as old as the Eocene; Coral Sea plankton; diameter 30 μm.

Siliceous skeleton of the silicoflagellate *Distephanus speculum*; Sydney coastal waters; diameter 30 μm.

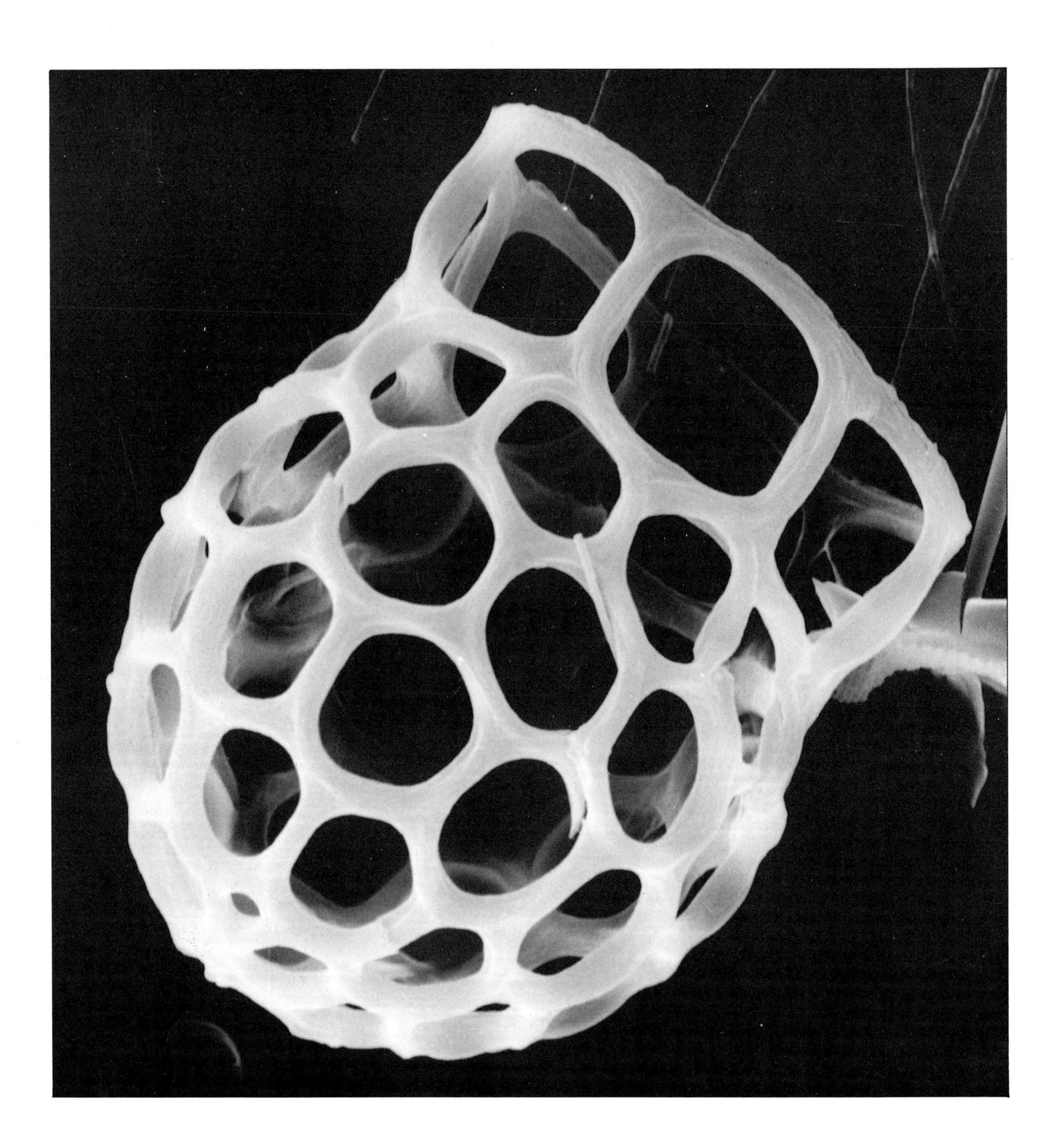

Tintinnid *Dictyocysta elegans,* a small planktonic ciliate that lives in a basket-like container (lorica). These animals feed with a crown of cilia protruding from the open end of the lorica; Coral Sea; diameter 50 μm.

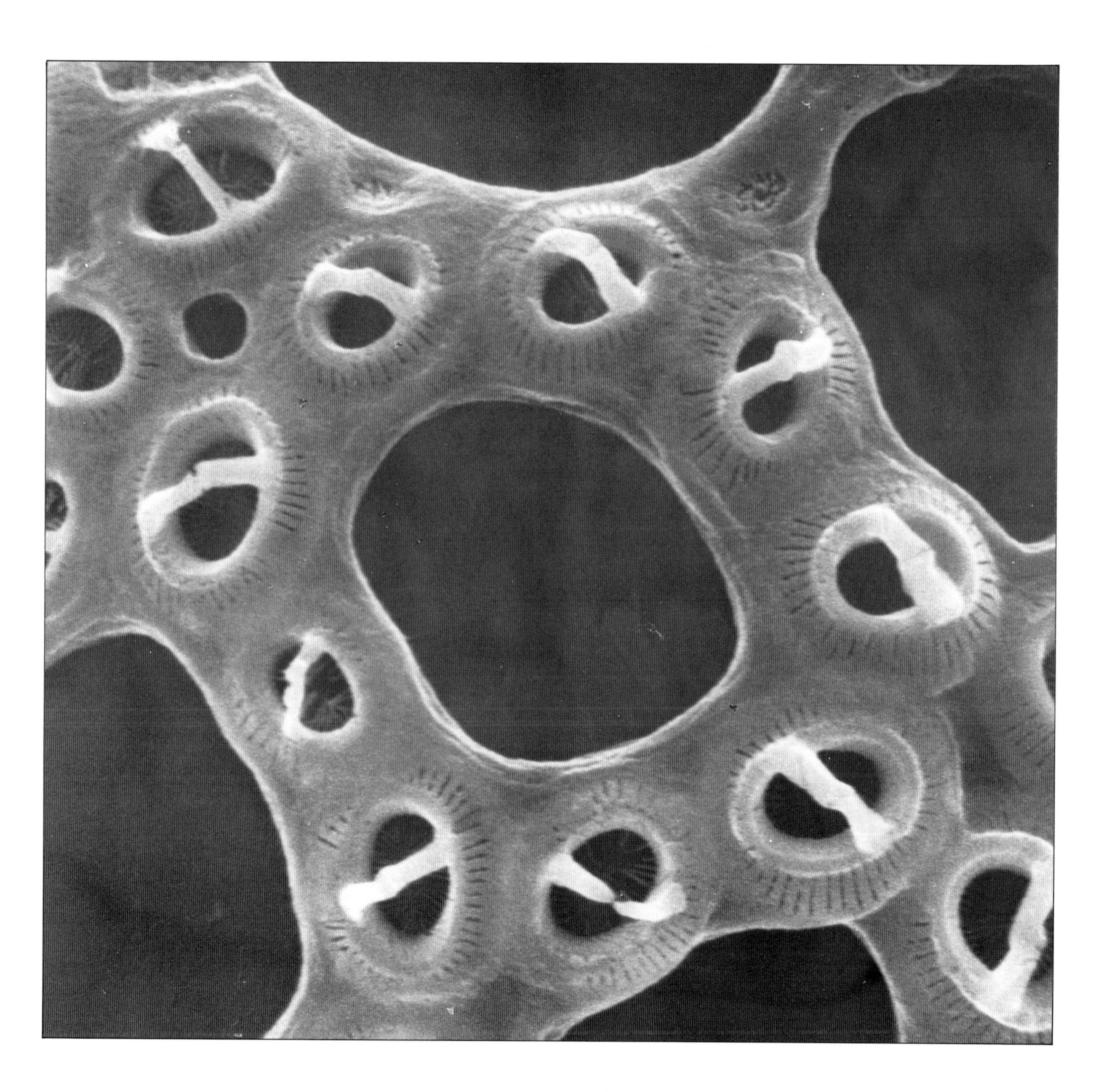

Detail of another *Dictyocysta* species, which has decorated its lorica with the scales of the coccolithophorid *Gephyrocapsa oceanica* (compare Plate 11); Coral Sea.

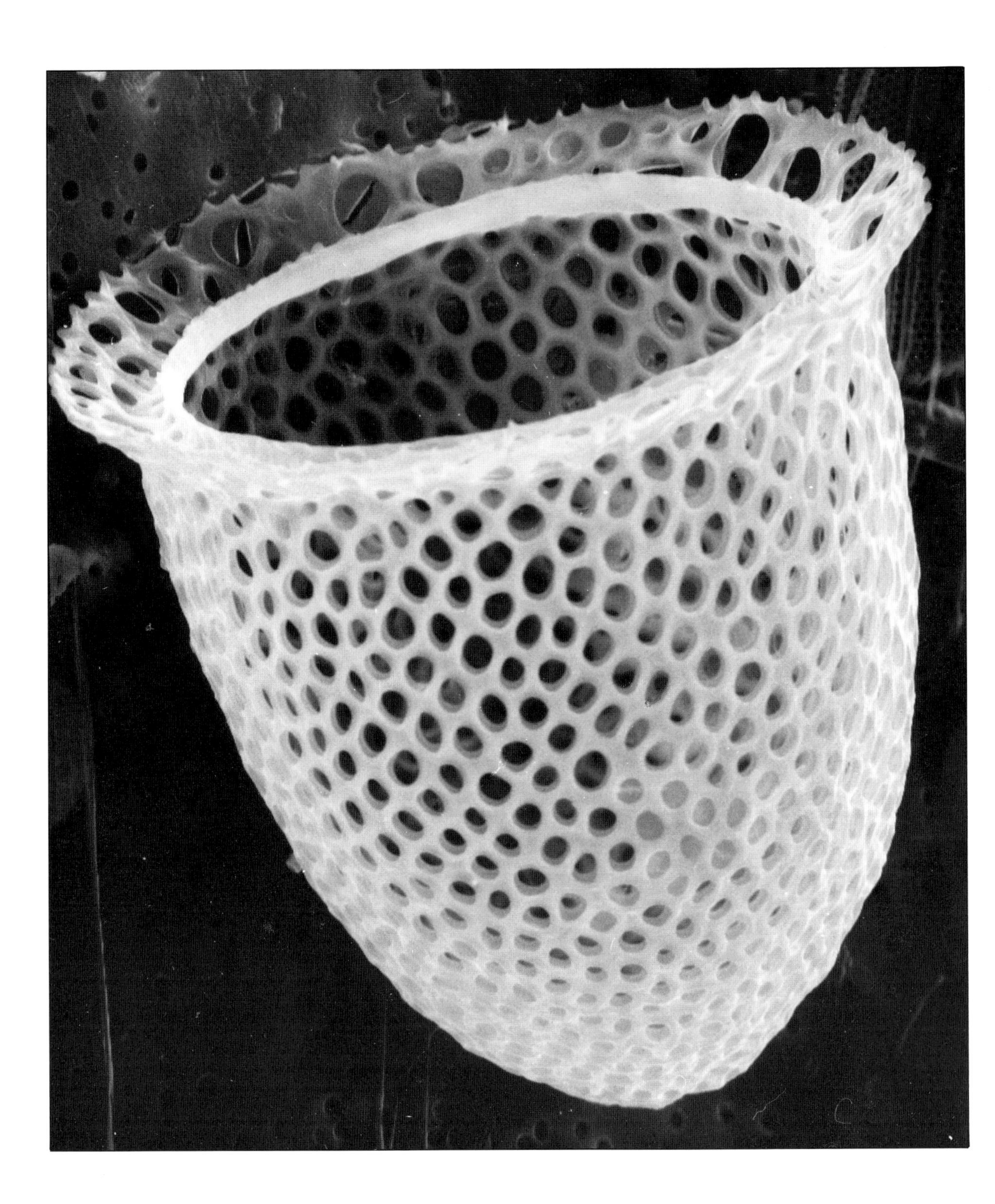

Chitinoid lorica of the tintinnid *Cyttarocylis*; Coral Sea; diameter 70 μm.

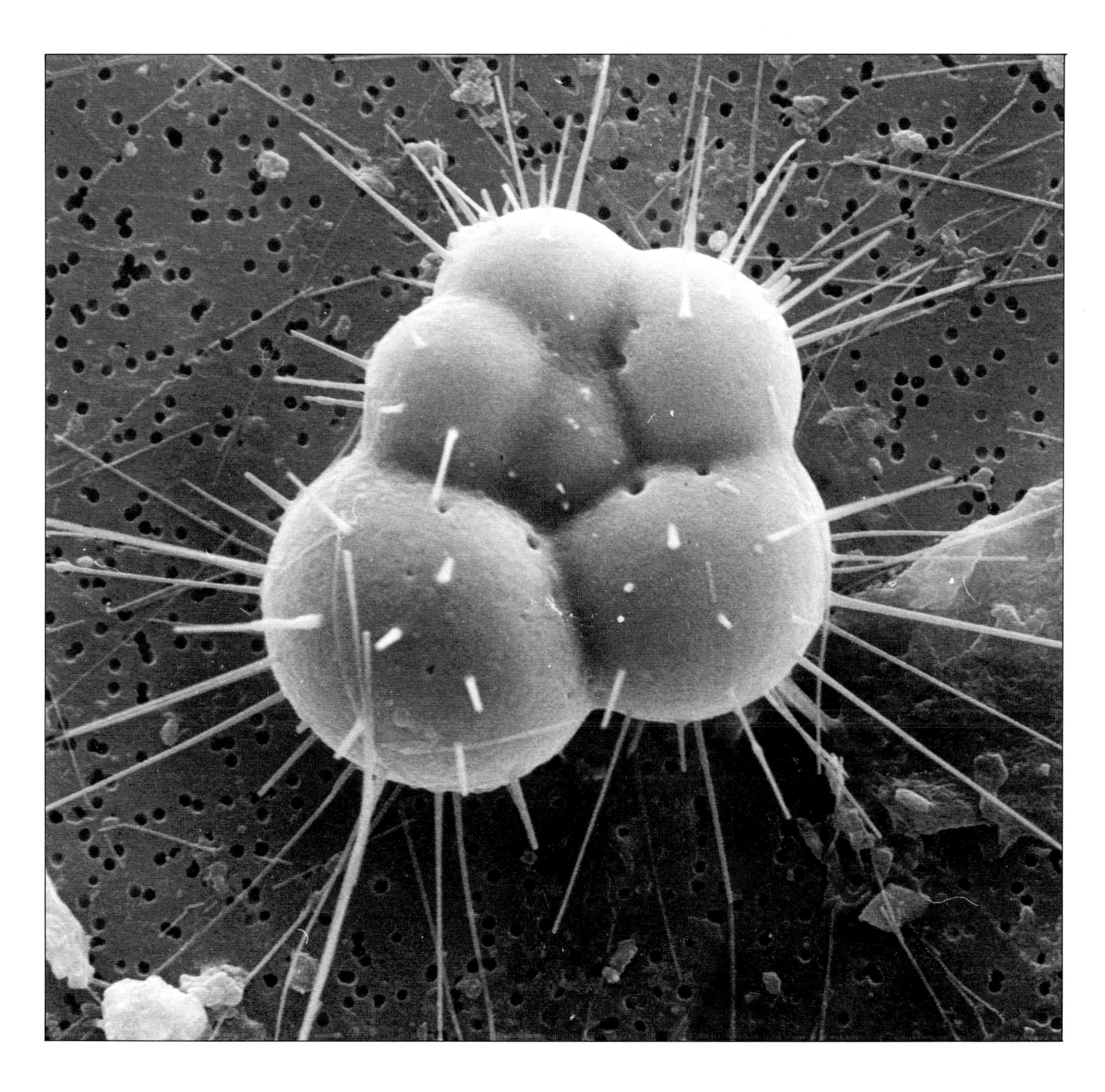

Calcareous shell of the foraminiferan *Globigerina*. These plankton animals extend pseudopods through small pores in search of food; Coral Sea; diameter 100 μm.

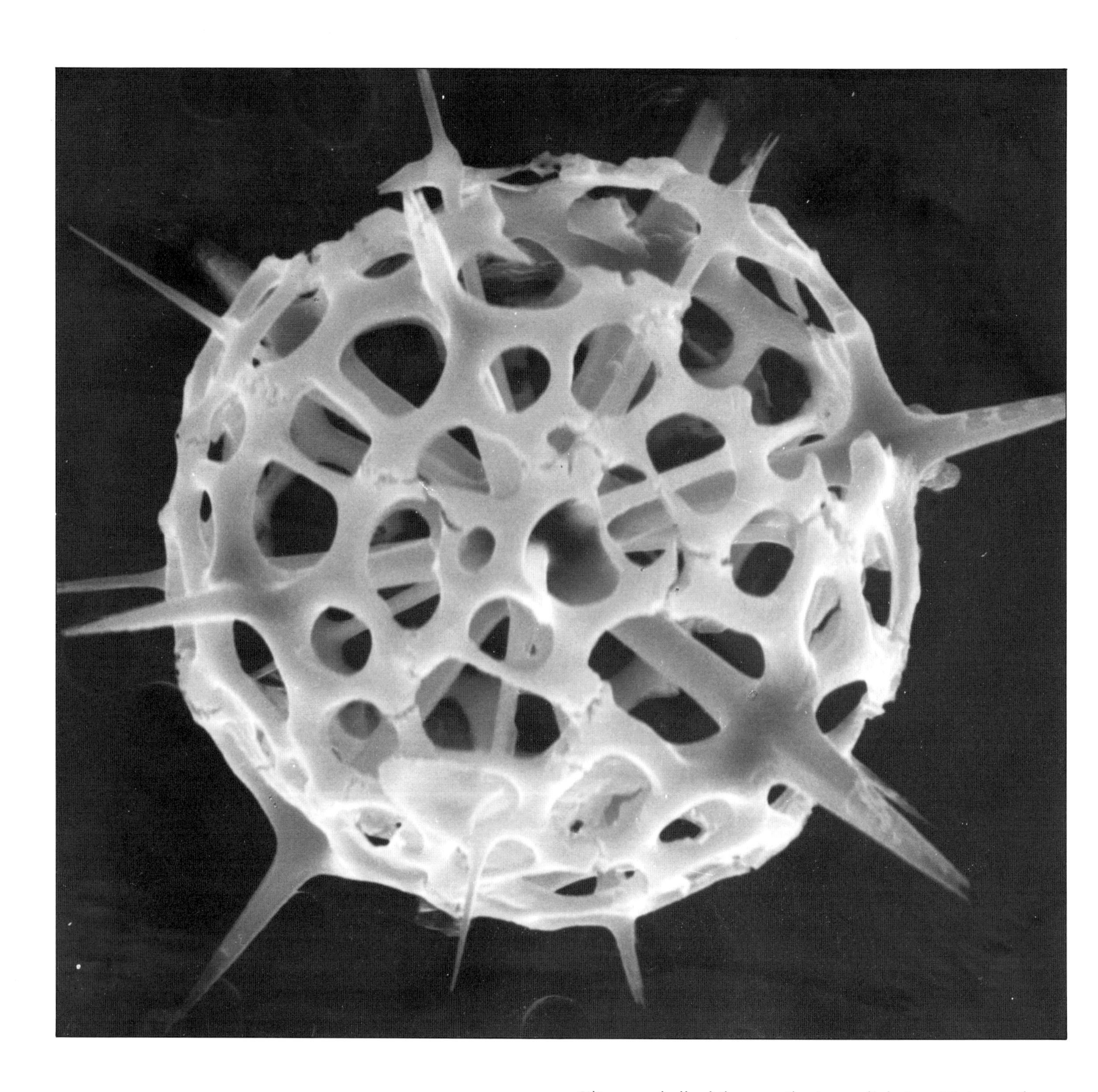

Siliceous shell of the acantharian radiolarian *Dictyacantha;* Coral Sea; diameter 100 μm.

Star-like acantharian *Trizona*; Coral Sea; diameter 100 μm.

SECTION 3

Acknowledgements

The author is indebted to Dr Shirley Jeffrey for encouragement and for her efforts to establish an 'in home' electron microscope facility at the CSIRO Marine Laboratories in Hobart, which made much of this work possible (1985-1988). Dr Tony Rees provided expert assistance in the new facility. Dr Maret Vesk and Ms Diane Hughes are thanked for training and hospitality at the Electron Microscope Unit of Sydney University during 1980 to 1984. Drs Vivienne Mawson and Shirley Jeffrey provided helpful comments on the manuscript, Mr Chris Bolch prepared the micrographs in pages 50 and 51, and Ms Isla MacGregor provided artistic encouragement.

Further Reading

BACH, K. and BURKHARDT, B. (eds), 1984. Shells in Nature and Technics. Diatoms I. Publications of the Institute for Lightweight Structure No. 28.

CLAYTON, M. and KING, R. (eds), 1981. *Marine Botany - an Australasian perspective*. Longman Cheshire, Melbourne (first edition; second edition will be published in 1989).

DODGE, J.D., 1985. Atlas of Dinoflagellates. Farrand Press, London, 128 pp.

HALLEGRAEFF, G.M., 1984. Coccolithophorids (calcareous nanoplankton) from Australian waters. *Botanica Marina* 27, 229-247.

HALLEGRAEFF, G.M., 1984. Species of the diatom genus *Thalassiosira* in Australian waters. *Botanica Marina* 27, 495-513.

HALLEGRAEFF, G.M., and JEFFREY, S.W., 1984. Tropical phytoplankton species and pigments of continental shelf waters of North and North-West Australia. *Marine Ecology Progress Series* 20, 59-74.

HALLEGRAEFF, G.M., and JEFFREY, S.W., 1985. Marine plankton pastures studied with the electron microscope. *Australian Science Magazine* 3, 20-24.

HALLEGRAEFF, G.M. and LUCAS, I.A.N., 1988. The marine dinoflagellate genus *Dinophysis* (Dinophyceae): photosynthetic, neritic and non-photosynthetic, oceanic species. *Phycologia* 27, 25-42.

HALLEGRAEFF, G.M., STEFFENSEN, D.A. and WETHERBEE, R., 1988. Three estuarine Australian dinoflagellates that can produce paralytic shellfish toxins. *Journal of Plankton Research* 10, 533-541.

TAYLOR, F.J.R., 1976. Dinoflagellates from the International Indian Ocean expedition. Bibliotheca Botanica 132, 234 pp.

Glossary

alga *(plural: algae)*
any of various primitive, chiefly aquatic, one-celled or multicellular plants that lack true stems, roots and leaves, and which usually contain chlorophyll. Included among the algae are large kelps and other seaweeds (*macro-algae*) and the microscopic diatoms, dinoflagellates and coccolithophorids (*micro-algae*)
antapex
bottom end of the dinoflagellate cell
apex
top end of the dinoflagellate cell
apical pore
distinctive pore-like structure, present on top of some dinoflagellate cells
areola
the regularly repeated perforations of the siliceous diatom cell wall; the regularly repeated circular depressions in the cellulose wall of dinoflagellates
armoured
dinoflagellate cells with a thick cellulose wall (*theca*) composed of rigid plates
bioluminescence
the emission of visible light by living organisms, such as the fireflies on the land and the dinoflagellates in the sea
blue-green algae
primitive single-celled algae, characterised by the absence of a nucleus and other membrane-bound organelles, and their (often) blue-green colour
chain
a group of micro-organisms which are connected in a linear arrangement
see *colony*
chlorophyll *a*
the green colouring substance found in all photosynthetic algae and higher plants
ciguatera
human illness caused by eating certain tropical coral reef fish which are contaminated with dinoflagellate toxins (e.g. from *Gambierdiscus toxicus*)
coccolithophorids
a class of single-celled golden-brown coloured flagellates, which are covered by calcareous scales (*coccoliths*)
coccolith
see coccolithophorid
colony
a group of the same kind of micro-organisms, which are connected and coexist in close association
cyst
a thick-walled stage in the life-history of a dinoflagellate, also called resting spore. The cyst wall is often made up of resistant *sporopollenin* and may be ornamented with spines. Cysts are often produced at the onset of adverse conditions, and when conditions are suitable, they may germinate to seed new plankton blooms
diarrheic shellfish poisoning
human illness caused by eating shellfish contaminated with certain dinoflagellate toxins (e.g. from *Dinophysis acuminata*)
diatoms
single-celled golden-brown coloured algae, having siliceous cell walls consisting of two overlapping symmetrical parts
dinoflagellates
single-celled golden-brown algae, characteristically having a large nucleus with clearly visible chromosomes, and with two flagella, protruding from the girdle groove and sulcus groove, respectively.
electron microscope
a microscope that uses electrons rather than light to produce magnified images, especially of objects having dimensions smaller than the wavelengths of visible light
epitheca
the upper half of the diatom or dinoflagellate cell
etymology
the semantic derivation of a scientific species name
flagellum (plural: *flagella*)
whip-like extensions of the cells of certain micro-organisms. Mainly used for locomotion
foraminiferans
single-celled protozoans with perforated calcareous shells through which numerous pseudopodia protrude
frustule
the silica cell wall of a diatom, composed of an upper part (*epivalve* or *epitheca*), lower part (*hypovalve* or *hypotheca*) and connecting bands (*girdle*)
girdle bands
the part of the diatom frustule between epitheca and hypotheca
girdle groove
the horizontal groove which encircles the dinoflagellate cell and which contains the transverse flagellum
girdle view
the diatom cell in lateral view; see *valve view*
green flagellates
single-celled motile plankton algae, characterized by the presence of the olive-green pigment chlorophyll *b*
horn
extension of the cellulose plates of a dinoflagellate cell
hypotheca
the lower half of the diatom or dinoflagellate cell; see *epitheca*
labiate process
hollow tube-like projection or opening through the diatom cell wall, which is supported internally by a flattened tube or longitudinal slit, often surrounded by two lips
light microscope
an optical instrument that uses a combination of glass lenses to produce magnified images of objects too small to be seen by the unaided human eye
list
extension of the cellulose plates of dinoflagellate cells, usually around the girdle groove

micro-algae
see alga
micron
one thousandth of a millimetre *(μm)*
naked
plankton cells bounded by a membranous covering only. These organisms are very easily damaged beyond recognition.
nanoplankton
the smallest plankton species, in the size range 2 to 20 micron
neurotoxins
toxins that disrupt the normal functioning of the human nervous system
ocellus
eye-like structure on the diatom cell wall, consisting of a plate of silica, normally with a thickened rim and pierced by closely packed holes (e.g. in *Auliscus sculptus*
occluded process
hollow tube-like projection on the diatom cell which does not penetrate the cell wall; see *strutted process*
paralytic shellfish poisoning
human illness caused by eating shellfish which are contaminated with certain dinoflagellate toxins (e.g. from *Gymnodinium catenatum* or *Pyrodinium bahamense*). In extreme cases this can lead to death through respiratory paralysis
phytoplankton
plant plankton; see *plankton*
plankton
plant and animal organisms, mostly microscopic, that float or drift passively with the currents, both in fresh or salt water
plankton bloom
prolific growth of plankton algae; see *red tide*
pseudonodulus
circular hole near the margin of the diatom cell wall (e.g. in *Roperia tesselata*)
radiolarians
single-celled marine protozoans with rigid siliceous skeletons and spicules
red tide
dense concentrations of plankton organisms (mostly dinoflagellates) that can colour the sea red or brown
scanning electron microscope
a type of electron microscope in which a beam of electrons scans the outer surface of objects previously coated with a thin layer of gold or platinum
silicoflagellates
single-celled golden-brown coloured flagellate algae with characteristic internal siliceous skeleton
spine
solid projection on the surface of the diatom or dinoflagellate cell wall
strutted process
hollow tube-like projection of the diatom cell wall, supported at the base by struts
sulcus
the vertical groove on the dinoflagellate cell, extending onto epitheca and hypotheca, which guides the longitudinal flagellum
symbiosis
the living together of two species of organisms with mutual advantages for both
theca
the complete cell covering of the dinoflagellate cell
tintinnid
single-celled protozoans with a crown of cilia protruding from a basket-like lorica
transmission electron microscope
a type of electron microscope in which electrons pass through (sections of) thin biological specimens
valve view
the diatom cell viewed from the top or bottom; see *frustule*
zooplankton
animal plankton; see *plankton*

Index